Computer Numerical Control

SIMPLIFIED

Steve Krar   Arthur Gill Peter Smid

Industrial Press Inc.
New York

Library of Congress Cataloging-in-Publication Data

Krar, Stephen F.
    Computer numerical control simplified/Steve Krar, Arthur Gill, Peter Smid.
    p. cm
  ISBN 0-8311-3133-0
    1. Machine-tools—Numerical control. I. Gill, Arthur, 1930– II. Smid, Peter. III. Title

TJ1189.K745 2000
621.9′023—dc21   00-056691

Industrial Press Inc.
200 Madison Avenue
New York, NY 10016-4078

First Edition

Computer Numerical Control Simplified

1 2 3 4 5 6 7 8 9 10

Contents

Preface

The term numerical control is a widely accepted and commonly used term in the machine tool industry. Numerical control (NC) enables an operator to communicate with machine tools through a series of numbers and symbols.

NC and computer numerical control (CNC) have brought tremendous changes to the metalworking industry. New machine tools in combination with NC or CNC have enabled industry to produce parts consistently to accuracy undreamed of only a few years ago. The same part can be reproduced to the same degree of accuracy any number of times if the NC tape has been properly prepared or if the computer has been properly programmed. The operating commands, which control the machine tool, are executed automatically with amazing speed, accuracy, efficiency, and repeatability.

The ever-increasing use of NC and CNC in industry has created a need for personnel who are knowledgeable about and capable of preparing the programs that guide the machine tools to produce parts to the required shape and accuracy. With this in mind, the authors have prepared this textbook to take the mystery out of CNC—to put it into a logical sequence and express it in simple language that everyone can understand.

The main features of this book are:

1. It is written in a clear concise language that most students easily understand.
2. The learning objectives are outlined for each unit or section.
3. Key words are listed that will be explained in the unit or section.
4. The numerous photos and illustrations are used to draw attention on specific items or areas to make learning easier and more effective.
5. The basics of the Cartesian coordinate system, part system, and the machine coordinate system are explained with clear, concise examples and illustrations.
6. The absolute and incremental dimensioning systems and their use are covered along with illustrated examples of each.
7. Knowledge review questions at the end of each unit are designed to act as a review and stimulate original thinking.
8. The book covers both bench-top teaching size and industrial CNC turning centers (lathes) and CNC machining centers (mills).
9. Each programming code and step is fully explained so that the learner quickly becomes familiar with CNC programming language.
10. A sample part for both CNC lathe and the CNC mill is completely programmed and each program code is explained as it is being used.

11. An introduction to the fundamentals of CAD/CAM Mill and Lathe programming is explained using visuals for the user to check for accuracy.
12. The glossary explains the key technical words and terms used in CNC.
13. The appendix contains useful tables and formulas used in calculations for machining and turning centers.

It is advisable for a person who wishes to follow a career in CNC to have some background in machining fundamentals, in order to know how to program the sequence of operations required to produce a finished part. An introductory machine trades course should provide a student with a knowledge of the selection of the proper machine tool, machining operations, cutting tools, and cutting technology required to machine the part. It would be very difficult for a person to become a good CNC programmer or machine operator without this knowledge.

CNC plays an important role in manufacturing along with VNC (Virtual Numerical Control), CAD (Computer-Aided Design), CAM (Computer-Aided Manufacturing), CIM (Computer-Integrated Manufacturing), FMS (Flexible Manufacturing Systems), and Virtual and Digital Manufacturing. The advances in microelectronics and computers are truly revolutionizing manufacturing.

Steve Krar
Arthur Gill
Peter Smid

Acknowledgments

The authors wish to express their sincere appreciation to Alice Krar for the many hours spent typing, proofreading, and checking the manuscript. Her assistance was of prime importance in making the book as clear as possible for the learner and teacher alike. A special thanks to Jonathan Gill for his assistance in scanning manuscript, recording artwork, and general assistance in producing the final manuscript.

Many thanks for the professional assistance of the Industrial Press team, John Carleo, Editor/Marketing Director, and Janet Romano, Art Director/Production Manager who so capably guided this book through its many stages.

We owe a special debt of gratitude to the many students, teachers, and industrial personnel who reviewed sections of the book and offered constructive criticism and suggestions for improving this text. We greatly appreciate the assistance of the following firms, which were kind enough to review sections of the manuscript and supply technical information and illustrations for this book.

Allen-Bradley Company
Bausch & Lomb
Bridgeport Div. Textron, Inc.
Butterfield Division, Union Twist Drill
Carboloy, Inc.
Cincinnati Milacron, Inc.
Cleveland Twist Drill Company
Coleman Engineering Company, Inc.
Deckel-Maho, Inc.
Deneb Robotics, Inc.
Denford, Inc.
Electronic Industries Association
Fadal Engineering, Inc.
Forkardt, Inc.
Greenfield Tap and Die Company
Haas Automation, Inc.
Hertel Carbide Ltd.
Hewlett Packard

Hitachi-Seiki U.S.A. Inc.
Icon Corporation
Ingersol Milling Machine Company
Kennametal Inc.
Kelmar Associates
Makino, Inc.
Mazak Corporation
Modern Machine Shop
Moore Special Tool Company
Northwestern Tools, Inc.
Numeridex Inc.
Philips Electronic Instruments
PowerHold Inc.
Rockwell International
Taft-Peirce Mfg. Company
Weldon Tool Company
J.H. Williams & Company

Authors

Steve F. Krar spent 15 years in the machine tool trade and later graduated from the Ontario College of Education, University of Toronto, with a Type A Specialist's Certificate in Technical Education. After 20 years of teaching, he devoted full time to researching and coauthoring over 50 books on machine tools and manufacturing technology. His *Technology of Machine Tools*, now in its fifth edition, is recognized as one of the leading texts in the world on the subject; it has been translated into four languages. During his years of research, he has studied under Dr. W. Edwards Deming and has been associated with GE Superabrasives, besides countless other leading machine tool manufacturers. He was invited twice to China to teach and share his knowledge about modern machining and manufacturing technology. A former Associate Director of the GE Superabrasives Partnership for Manufacturing Productivity, Steve Krar is a life member of the Society of Manufacturing Engineers.

Arthur R. Gill served an apprenticeship as a tool and die maker. After 10 years in the trade, he entered the Ontario Community College system as a professor and coordinator of precision metal trades and apprenticeship training. During his 30 years at Niagara College in St. Catharines, he has been a member of the Ontario Precision Metal Trades college curriculum committee for apprenticeship training and head of Apprenticeship for Ontario. Art is a member of the Society of Manufacturing Engineers and has worked closely with industry to continually improve manufacturing technology. He has coauthored *CNC Simplified* and *Computer Numerical Control Programming Basics* with Steve Krar. In 1991 he was invited by the People's Republic of China to assist in developing a Precision Machining and Computer Numerical Control (CNC) training facility at Yueyang University in Hunan Province.

Peter Smid graduated from high school in Czechoslovakia with a specialty in machine shop training. After graduation, he entered industry, completed an apprenticeship program, and gained valuable experience as a machinist skilled on all types of machine tools. Peter emigrated to Canada in 1968 and spent the next 26 years employed in the machine tool industry as a machinist and a tool and die maker.

In the early 1970s he became involved in Computer Numerical Control (CNC) as a programmer/operator and devoted the next two decades to becoming proficient in all aspects of computerized manufacturing. In 1989 he became an independent consultant and hundreds of companies have used his CNC and CAD/CAM skills to improve their manufacturing operations. While working as a consultant, Peter found time to write *CNC Programming Handbook*, an extraordinarily comprehensive guide to the subject.

SECTION 1

Introduction to CNC

Unit 1 Measurement Fundamentals

Unit 2 CNC Machines

INTRODUCTION TO CNC

Numerical control (NC) is the operation of a machine tool by a series of coded instructions consisting of numbers, letters of the alphabet, and symbols that the machine control unit (MCU) can understand. These instructions are converted into electrical pulses of current that the machine's motors and controls follow to carry out machining operations on a workpiece. The numbers, letters, and symbols are coded instructions that refer to specific distances, positions, functions, or motions, which the machine tool can understand as it machines the workpiece.

HISTORY

A form of NC was used in the early days of the industrial revolution, as early as 1725, when knitting machines in England used punched cards to form various patterns in cloth. Even earlier than this, rotating drums with prepositioned pins were used to control the chimes in European cathedrals and some American churches. In 1863, the first player piano was patented; it used punched paper rolls, through which air passed to control the order in which the keys were played automatically, Fig. 1–2.

The principle of mass production (interchangeable manufacture), developed by Eli Whitney, transferred many operations and functions originally performed by skilled artisans to the machine tool. As better and more precise machine tools were developed, the system of interchangeable manufacture was quickly adopted by industry in order to produce large quantities of identical parts. In the second half of the nineteenth century, a wide range of machine tools was developed for the

Fig. 1-1 Machine tool technology in action. *(Haas Automation, Inc.)*

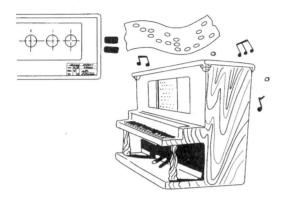

Fig. 1-2 The player piano used air passing through holes in a drum to operate the keys in a programmed sequence. *(Coleman Engineering Company)*

basic metal–cutting operations, such as turning, drilling, milling, and grinding. As better hydraulic, pneumatic, and electronic controls were developed, better control over the movement of machine slides became possible.

In 1947, the U.S. Air Force found that the complex designs and shapes of aircraft parts such as helicopter rotor blades and missile components were causing problems for manufacturers, who could not keep up projected production schedules. At this time, John Parsons, of the Parsons Corporation, of Traverse City, Michigan, began experimenting with the idea of making a machine tool generate a "thru-axis curve" by using numerical data to control the machine tool motions. In 1949, the U.S. Air Material Command awarded Parsons a contract to develop NC and in turn speed up production. Parsons subcontracted this study to the Servomechanism Laboratory of the Massachusetts Institute of Technology (MIT), which in 1952 successfully demonstrated a vertical spindle Cincinnati Hydrotel, which made parts through simultaneous three-axis cutting tool movements. In a very short period of time, almost all machine tool manufacturers were producing machines with NC.

At the 1960 Machine Tool Show in Chicago, over a hundred NC machines were displayed. Most of these machines had relatively simple point-to-point positioning, but the principle of

NC was now firmly established. From this point, NC improved rapidly as the electronics industry developed new products. At first, miniature electronic tubes were developed, but the controls were big, bulky, and not very reliable. Then solid-state circuitry and, eventually, modular, or integrated, circuits were developed. The control unit became smaller, more reliable, and less expensive. The development of even better machine tools and control units helped spread the use of NC from the machine tool industry to all facets of manufacturing.

NC EVOLVES INTO CNC

The introduction of software-based controls in the early 1970s replaced the NC hardware design with complete computer logic that had more capacity and could be programmed for a variety of functions at any time, Fig. 1-3. This made it possible to revise, modify, or update CNC programs or parts of programs at any time on a computer. In turn, CNC machines with their menu-selected displays, advanced graphics, and ease of programming became easier to use.

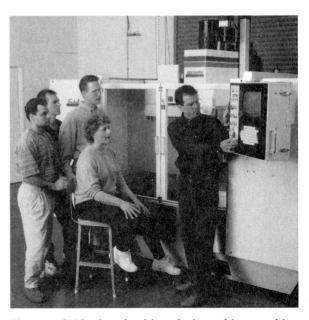

Fig. 1-3 CNC played a big role in making machine tools more accurate and greatly increased their productivity. *(Fadal Engineering, Inc.)*

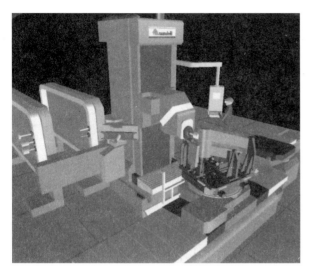

Fig. 1-4 Virtual NC makes it possible to check out the accuracy of a CNC program and the machining operation on a computer screen. *(Deneb Robotics, Inc.)*

VIRTUAL NC

The latest in the development of CNC is Deneb's Virtual NC®, a superior interactive 3D simulation setting, that allows a person to visualize and analyse the operation of a machine tool, its CNC controller, and the material–removal process, Fig. 1–4. It allows a user to improve the quality of CNC programs, eliminate tool and machine crashes due to program errors, and produce the most efficient machining process. The result is maximized machine tool productivity.

A few of the advantages of Virtual NC are:

- **Rapid Modeling**—Attachments, tool changers, fixtures, etc., can be added to produce a realistic manufacturing process.
- **Training**—New operators can be trained without damaging the machine or tooling.
- **Cycle Information**—Material removal and machining data can be continuosly monitored.
- **Collision Avoidance**—Virtual NC automatically detects near misses and collisions, and stops the cycle, at the same time noting the program block in which correction is needed.
- **Concurrent Engineering**—Manufacturing engineers and programmers can evaluate the machine controller and the machining process without taking up valuable machine time or risking damage to the machine.

Unit (1)
Measurement Fundamentals

NC data processing (with numbers, letters, and symbols) is done in a computer or machine control unit (MCU) by adding, subtracting, multiplying, dividing, and comparing. The computer can be programmed to recognize an A command before a B command, an item 1 before an item 2, or any other elements in their sequential order. Subtracting is done by adding negative values, multiplying is done through a series of additions, and dividing is done by a series of subtractions. The computer is capable of handling numbers very quickly; the addition of two simple numbers may take only one billionth of a second (a nanosecond).

OBJECTIVES

After completing this unit you should be able to:

1. Know and understand the Cartesian coordinate measuring system.
2. Identify three of the more common machine tools using CNC.
3. Know the difference between the absolute and incremental systems of measurement.

KEY TERMS

absolute	decimal system
binary numbers	incremental
Cartesian coordinate system	nanosecond

INCH/METRIC DIMENSIONING

In an attempt to standardize dimensioning throughout the world, the United States has been working with European Standards organizations for many years to arrive at an agreeable solution. As early as 1982 the American Standards publication ANSI Y14.5M—1982 and reaffirmed in the publication ASME Y14.4M—1994, the following has been adopted as the standard. This standard is used throughout the book in the drawings and the text.

1. **Decimal Inch Dimensioning (See Fig. 1-1-1)**
   • Where the dimension is less than one inch, a *zero is not used* before the decimal point.

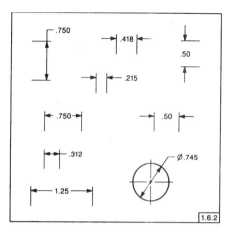

Fig. 1-1-1 The ASME recommended use of decimal inch dimensions.

- Every dimension should be expressed to the same accuracy as its tolerance. Zeros should be added to the right of the decimal point wherever necessary.
2. **Metric (Millimeter) Dimensioning (See Fig. 1-1-2)**
- Where a dimension is less than one millimeter, *a zero precedes* the decimal point.
- Where the number is a whole number, no zero or decimal point is shown.
- Where the dimension is larger than the whole number by a decimal fraction of one millimeter, the last digit to the right of the decimal point is not followed by a zero.

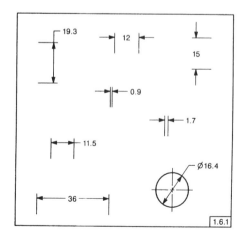

Fig. 1-1-2 The ASME recommended use of metric (millimeter) dimensions.

BINARY NUMBERS

Primitive people used their 10 fingers and 10 toes to count numbers and from this evolved our present Arabic, or decimal system where "base ten," or the power of 10, is used to signify a numerical value. Computers and MCUs, in contrast, use the binary or base 2 system to recognize numerical values. Knowledge of the binary system is not essential for the programmer or operator since both the computer and the MCU can recognize standard decimal system and convert it to binary data.

To provide an understanding of the binary system, let us compare it with the decimal system.

1. **Decimal System** In the decimal system the value of each digit depends on where it is placed in relation to the other digits in a number. The number one (1) by itself is worth 1, but if it is placed to the left of one or two zeros (0), it is worth 10 and 100, respectively. Before numbers can be added or subtracted, they must first be arranged in their proper place columns. In the decimal system, each position to the left of a decimal point means an increase in the power of 10.

2. **Binary System** The binary system uses only two digits zero (0) and one (1), and is based on the power of two (2) instead of ten (10), as in the decimal system. Each position to the left means an increase by the power of two (2). For example:

$$2^1 = 2$$
$$2^2 = 4$$
$$2^3 = 8 \ (2 \times 2 \times 2)$$
$$2^4 = 16 \ (2 \times 2 \times 2 \times 2)$$
$$2^5 = 32$$

Therefore, any numerical value can be shown using only the two digits, one (1) and zero (0). Since there are only two digits, the binary system is often called the "ON-or-OFF" system. For example:

$$1 = ON \qquad 0 = OFF$$

On numerical punched tape, a hole represents a one (1) and no hole represents a zero (0).

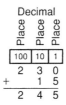

Decimal

Place	Place	Place
100	10	1
2	3	0
1		5
2	4	5

(+ before the second addend)

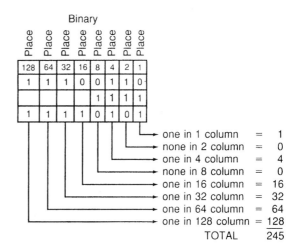

Binary

Place	Place	Place	Place	Place	Place	Place	Place
128	64	32	16	8	4	2	1
1	1	1	0	0	1	1	0
				1	1	1	1
1	1	1	1	0	1	0	1

one in 1 column = 1
none in 2 column = 0
one in 4 column = 4
none in 8 column = 0
one in 16 column = 16
one in 32 column = 32
one in 64 column = 64
one in 128 column = 128
TOTAL   245

Table 1-1-1 Comparison of the Decimal and Binary Systems

Table 1-1-1 illustrates the differences between the decimal and binary systems; the example is used to clarify both systems. Binary numbers are important for the computer and the MCU, which process information at a very high speed. Binary notations are used because electrical circuits are stable in either of two conditions: ON (POSITIVE or CHARGED) or OFF (NEGATIVE or DISCHARGED). Many control (MCU) settings can be set to ON-OFF, OPEN-CLOSED, IN-OUT and similar selection of any two options.

CARTESIAN COORDINATE SYSTEM

Almost everything that can be produced on a conventional machine tool can be produced on a computer numerical control machine tool, with its many advantages. The machine tool movements used in producing a product are of two basic types: point-to-point (straight-line movements) and continuous path (contouring movements), Fig. 1-1-3A and B.

The French mathematician and philosopher Rene Descartes devised the Cartesian or rectangular, coordinate system. With this system, any specific point can be described in mathematical terms from any other point along three perpendicular axes. This concept fits machine tools perfectly since their construction is generally based on three axes of motion (X, Y, Z) plus an axis of rotation. On a plain vertical milling machine, the X axis is the horizontal movement (right or left) of the table, the Y axis is the table cross movement (toward or away from the column), and the Z axis is the vertical movement of the knee or the spin-

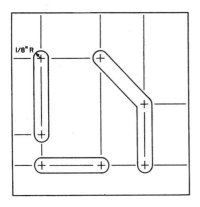

A

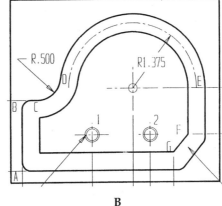

B

Fig. 1-1-3 Point–to–point or straight–line movement and continuous path or contour movement. *(Allen–Bradley)*

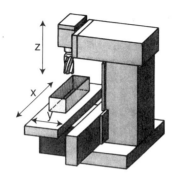

Fig. 1-1-4 The axes and directions of axes when work- ing with a CNC vertical spindle machine. *(Deckel–Maho, Inc.)*

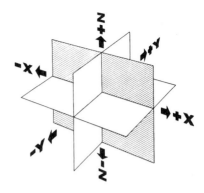

Fig. 1-1-6 The three-dimensional coordinate planes (axes) used in NC. *(Superior Electric Company)*

dle, Fig. 1-1-4. CNC systems use rectangular co- ordinates because the programmer can locate every point on a job precisely and independently from each other.

When points are located on a workpiece, two straight intersecting lines, one vertical and one horizontal at right angles to each other are used. Where these lines cross is called the origin or zero point, Fig. 1-1-5.

The three-dimensional coordinate planes are shown in Fig. 1-1-6. The X and Y planes (axes) are horizontal and represent horizontal machine table motions. The Z plane or axis represents the vertical tool motion. The Plus (+) and minus (–) signs indicate the direction from the zero point (origin) along the axis of movement. The four

quadrants formed when the X Y axes cross are numbered in a counterclockwise direction, Fig. 1-1-7. All positions in quadrant 1 are X positive (X+) and Y positive (Y+); the second quadrant, all positions are X negative (X–) and Y positive (Y+). In the third quadrant, all locations are X negative (X–) and Y negative (Y–); the fourth quadrant, all locations are X positive (X+) and Y negative (Y–).

In Fig. 1-1-7, point A would be 2 units to the right of the Y axis and 2 units above the X axis; assume that each unit equals 1 in. The location of point A would be X+ 2.000 and Y+ 2.000. For point B, the location would be X+ 1.000 and Y– 2.000. In CNC programming it is not necessary to

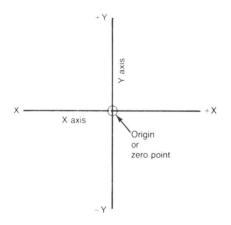

Fig. 1-1-5 Intersecting lines form right angles and es- tablish the zero point. *(Allen–Bradley)*

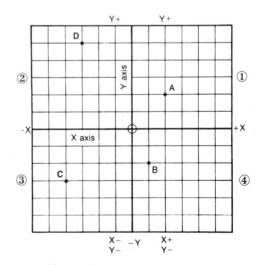

Fig. 1-1-7 The quadrants formed when the X and Y axes cross are used to locate points accurately from the XY zero, or origin, point. *(Allen–Bradley)*

indicate plus (+) values since these are assumed. However, the minus (–) values must be indicated and their omission makes the value positive by default. For example, the locations of both A and B would be indicated as follows:

A X2.000 Y2.000
B X1.000 Y–2.000

GUIDELINES

Since CNC is so dependent upon the system of rectangular coordinates, it is important to follow some guidelines. In this way, everyone involved in the manufacture of a part, the engineer, draftsperson, programmer, and machine operator, will understand exactly what is required.

1. Use reference points on the part itself, if possible. This makes it easier for quality control to check the accuracy of the part later.
2. Use Cartesian coordinates, specifying X, Y, and Z axes, to define all part surfaces.
3. Establish reference planes along part surfaces that are parallel to the machine axes.
4. Establish the allowable part tolerances at the design stage.
5. Describe the part so that the cutter path may be easily understood and programmed.

6. Dimension the part so that it is easy to understand its shape without calculations or guessing.

SUMMARY

- CNC programming consists of entering codes (numbers, letters, symbols, etc.) into a computer program that can be understand by the MCU (machine control unit).
- The binary system of measurement uses only two digits, zero(0) and one(1), and in the decimal system each position to the left of the decimal point increases the value by the power of 10.
- In the Cartesian coordinate system any point on a job or part can be located in mathematical terms along three perpendicular axes.
- On a CNC vertical spindle machine the X axis moves the machine table horizontally right or left, the Y axis moves the table towards or away from the column, and the Z axis moves the spindle vertically up or down.
- When programming, it is wise to define all parts using the X, Y, Z axes planes, establish reference planes parallel to the machine axes, and describe and dimension a part so that its properties can be understood easily.

KNOWLEDGE REVIEW

Measurement Fundamentals

1. Compare the decimal system with the binary system.
2. Why are only two numbers used in the binary system, and how do they affect electrical circuits?

Cartesian Coordinate System

3. Name the two basic classifications of positioning.

4. Why does the Cartesian coordinate system fit machine tools perfectly?

5. Using a vertical milling machine, define the X, Y, and Z axes.

6. Make a sketch of the four quadrants and explain what each represents.

7. Why is it important in CNC work that some guidelines be followed with the system of rectangular coordinates?

Unit 2
CNC Machines

Early machine tools were designed so that the operator stood in front of the machine while operating the controls. This design is no longer necessary, because in CNC the operator no longer controls the machine tool movements. On conventional machine tools, only about 20 percent of the time is spent removing material. With the addition of electronic controls, actual time spent removing metal has increased to 80 percent and even higher. Numerical control has also reduced the amount of time required to bring the cutting tool into each machining position.

OBJECTIVES

After completing this unit you should be able to:

1. Understand and state the purpose of two types of CNC machine tools.
2. Be familiar with the coordinate system and the functions of the X, Y, Z axes on various types of machines.
3. Know the principle regarding the use of the incremental and absolute programming systems.
4. State five main advantages of the use of CNC on machine tools.

KEY TERMS

absolute system	positioning systems
contouring	programming systems
incremental system	reference center
machining center	turning center

MACHINE TYPES

In the past, machine tools were kept as simple as possible in order to keep their costs down. Because of the ever-rising cost of labor, better machine tools, complete with electronic controls, were developed so that industry could produce more and better goods at prices that are competitive in the world marketplace.

CNC is being used on all types of machine tools, from the simplest to the most complex. The most common machine tools are single-spindle drilling machines, lathes, milling machines, turning centers, and machining centers.

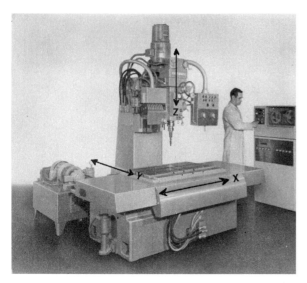

Fig. 1-2-1 A numerically controlled single-spindle drilling machine showing the X, Y, and Z axes. *(Cincinnati Machine, a UNOVA Co.)*

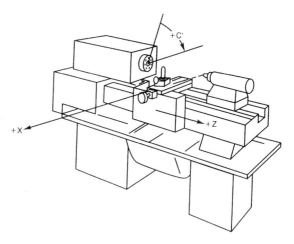

Fig. 1-2-2 The engine lathe cutting tool moves only on the X and Z axes. *(Electronic Industries Association)*

1. **Single-spindle drilling machine**—One of the simplest numerically controlled machine tools is the *single-spindle drilling machine*, Fig. 1-2-1.

Most drilling machines are programmed on three axes:

- The X axis controls the table movement to the right and left.
- The Y axis controls the table movement toward or away from the column.
- The Z axis controls the up or down movement of the spindle to drill holes to depth.

2. **Lathe** The *engine lathe*, one of the most productive machine tools, has always been a very efficient means of producing round parts, Fig. 1-2-2. Most lathes are programmed on two axes:

- The X axis controls the cross motion (in or out) of the cutting tool, (diameter control).
- The Z axis controls the carriage travel toward or away from the headstock (length control).

3. **Milling Machine** The milling machine has always been one of the most versatile machine tools used in industry, Fig. 1-2-3. Operations such as milling, contouring, gear cutting, drilling, boring, and reaming are only a few of the

many operations that can be performed on a milling machine. The milling machine can be programmed on three main axes:

- The X axis controls the table movement left or right.
- The Y axis controls the table movement toward or away from the column.
- The Z axis controls the vertical (up or down) movement of the knee or spindle.

4. **Turning Center** Turning centers were developed in the mid-1960s after studies showed that about 40 percent of all metal-cutting operations

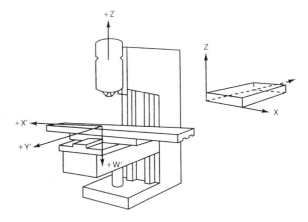

Fig. 1-2-3 The vertical knee and column milling machine operates on the X, Y, and Z axes. *(Electronic Industries Association)*

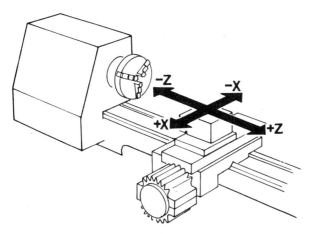

Fig. 1-2-4 Turning centers have two main axes; the X axis (tool cross movement) and Z axis (logitudinal movement) *(Emco Maier Corp.)*

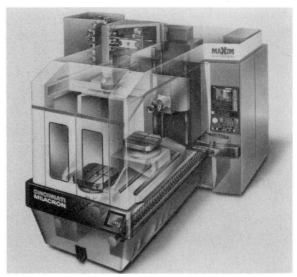

Fig. 1-2-5 Machining centers are capable of performing a variety of machining operations on a workpiece. *(Cincinnati Machine, aUNOVA Co.)*

were performed on lathes. These numerically controlled machines are capable of greater accuracy and higher production rates than were possible on the engine lathe. The basic turning center operates on only two axes, Fig. 1-2-4.

- The X axis controls the cross motion of the cutting tool (diameter control).
- The Z axis controls the lengthwise travel toward or away from the headstock (length control).

5. **Machining Center** Machining centers, Fig. 1-2-5, were developed in the 1960s so that parts did not have to be moved from machine to machine in order to perform various operations. These machines increased production rates because more operations could be performed on a workpiece in one setup. There are two main types of machining centers, the *horizontal* and the *vertical* spindle types.

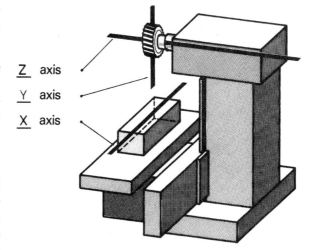

Fig. 1-2-6 Horizontal spindle machining centers have three axes of motion. *(Deckel Maho, Inc.)*

1. The **horizontal spindle machining center**, Fig. 1-2-6, operates on three axes:

- The X axis controls the table movement left or right.
- The Y axis controls the vertical movement (up or down) of the spindle.
- The Z axis controls the horizontal movement (in or out) of the spindle.

2. The **vertical spindle machining center**, Fig. 1-2-7, operates on three axes:

- The X axis controls the table movement left or right.
- The Y axis controls the table movement toward or away from the column.
- The Z axis controls the vertical movement (up or down) of the spindle.

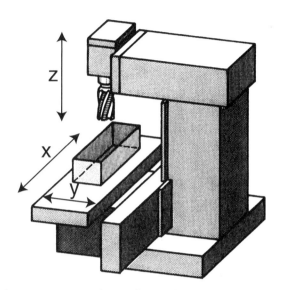

Fig. 1-2-7 Vertical spindle machining centers have three axes of motion. *(Deckel Maho, Inc.)*

PROGRAMMING SYSTEMS

Two types of programming modes, the absolute system and the incremental system, are used for CNC, Fig. 1-2-8. Both systems have applications in CNC programming, and no system is either right or wrong all the time. Controls on machine tools built today are capable of handling both absolute and incremental programming.

Absolute System

In the absolute system (G90), all dimensions or positions are given from a zero or reference point. For example, consider the person delivering newspapers using the absolute instructions with the street corner as the zero or reference point. The first house is 60 ft from the corner Fig. 1-2-9, the second house is 180 ft from the corner,

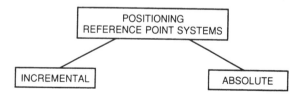

Fig. 1-2-8 Types of reference point positioning systems. *(Kelmar Associates)*

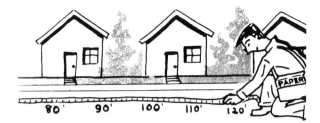

Fig. 1-2-9 The absolute system locates all houses requiring a newspaper by the distance each is from the street corner. *(Coleman Engineering Company)*

the third house is 240 ft from the corner, and so on. Therefore, all distances have been given from the corner, so that is the zero or reference point. In Fig. 1-2-10, the same workpiece is used as in Fig. 1-2-12, but all dimensions are given from the zero or reference point. Therefore, in the absolute system of dimensioning or programming, an error in any dimension is still an error, but the error is not carried on to any other location.

Absolute program locations are always given from a single fixed zero or origin point, Fig. 1-2-10. The zero or origin point may be a position on the machine table, such as the corner of the worktable or at any specific point on the workpiece. In absolute dimensioning and programming, each point or location on the workpiece is given as a certain distance from the zero or reference point, measured along an axis.

- A X plus (X+) command will cause the cutting tool to be located to the right of the zero or origin point.

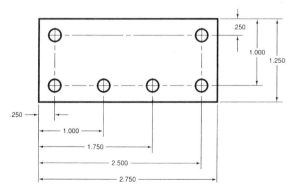

Fig. 1-2-10 A workpiece dimensioned in the absolute system mode. Note: All dimensions are given from a known point of reference. *(Icon Corporation)*

- A X minus (X–) command will cause the cutting tool to be located to the left of the zero or origin point.
- A Y plus (Y+) command will cause the cutting tool to be located toward the column.
- A Y minus (Y–) command will cause the cutting tool to be located away from the column.

In absolute programming, the G90 preparatory command indicates to the computer and MCU that the programming is to be in the absolute mode.

Incremental System

In the incremental system (G91), dimensions or positions are given from a previous known point. As an example of incremental instructions, consider a person who delivers newspapers to certain houses on a street. He or she could be given instructions to deliver a newspaper to the first house, 60 ft (feet) from the corner, Fig. 1-2-11. The second house that should have a paper could then be described as 120 ft from the first house, and the third house as 60 ft from the second. Note that all distances are expressed from the known previous point, in this case a house. Incremental dimensioning on a job print is shown in Fig. 1-2-12. One disadvantage of incremental positioning or programming is that if an error is made in any location, this error is carried over to all the locations made after this point.

Incremental program locations are always given as the distance and direction from the immediately preceding or last point, Fig. 1-2-12. Command codes that tell the machine to move

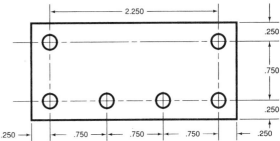

Fig. 1-2-12 A workpiece dimensioned in the incremental system mode. *(Icon Corporation)*

the table, spindle, and knee are here explained using a vertical milling machine as an example:

- A X plus (X+) command will cause the cutting tool to be located to the right of the last point.
- A X minus (X–) command will cause the cutting tool to be located to the left of the last point.
- A Y plus (Y+) command will cause the cutting tool to be located toward the column.
- A Y minus (Y–) will cause the cutting tool to be located away from the column.
- A Z plus (Z+) command will cause the cutting tool or spindle to move up or away from the workpiece.
- A Z minus (Z–) moves the cutting tool down or into the workpiece.

In incremental programming, the G91 preparatory command indicates to the computer and MCU that programming is to be in the incremental mode.

Fig. 1-2-11 The incremental system being used to locate the houses that require newspapers.
(Coleman Engineering Company)

POSITIONING (CONTROL) SYSTEMS

Positioning or control systems on CNC machines fall into two main categories, point-to-point and continuous path. Many variations of these two categories allow for a variety of milling-type operations, Fig. 1-2-13.

1. **Point-to-Point Positioning**
- Used to move the machine spindle rapidly, while above the work surface, to a specific location.

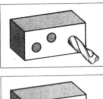

Point-to-point positioning control

Straight line control

- only milling parallel to the axis possible

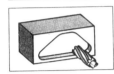

2 D-contouring control

- simultaneous milling in 2 axes

2$\frac{1}{2}$ D-contouring control

- 2D-milling in several planes
- rapid traverse in 3 axes

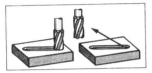

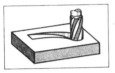

3 D-contouring control

- simultaneous milling in 3 axes

Fig. 1-2-13 Types of positioning (control) systems. *(Deckel–Maho, Inc.)*

- A machining operation is performed at the location and the spindle is retracted to a position above the work surface.
- The spindle is rapidly located at the next location where an operation must be performed.

2. **Straight Line Control**
- This control is used for any operations that involve straight-line machining parallel to the X or Y axis, and straight-line angular milling.

3. **2D—Contouring**
- This involves the simultaneous machining along two axes (X and Y); X and Z on lathes.

4. **2½ D—Contouring**
- This control allows the simultaneous machining along several planes.

5. **3D—Contouring**
- This involves the simultaneous machining along three axes, such as X, Y, Z.

6. Multi-axis contouring; 5-axes, 6+ axes.

FROM ART TO PART

The procedure for producing a part using CNC is very similar to the steps involved in producing a part on a conventional machine tool; the only difference being in the actual machining of the part.

- **In conventional machining**, the path that the cutting tool follows is controlled by the machine operator who turns the machine-slide handwheels.
  - The accuracy of the part depends largely on the skills that a machine operator has developed through many years of experience. It is very difficult for a human to produce two parts that are exactly the same.
- In CNC machining, the toolpath is controlled by the CNC program, which has been prepared from the information on the part drawing and related documentation.
  - The accuracy of the part depends upon the repeatability of the machine and the accuracy of the CNC program.
  - Once a program has been tested for accuracy, that each following part will be exactly the same as the first part with only minimum monitoring.

The basic steps of machining a part are outlined in Fig. 1-2-14.

1. **Reading the Technical Print or Drawing**
- A technical or engineering drawing should contain all the information required to plan the correct sequence of operations to produce an accurate part.
2. **Programming or Job Planning**
- **In conventional machining**, it is necessary to plan the correct sequence of operations to produce an accurate part.
- **In CNC machining**, the part program that will plan the sequence of operations and part sizes must be prepared.
3. **Inputting the Program**
- In CNC, the part program must be entered into a machine tool computer or an external computer (off-line).

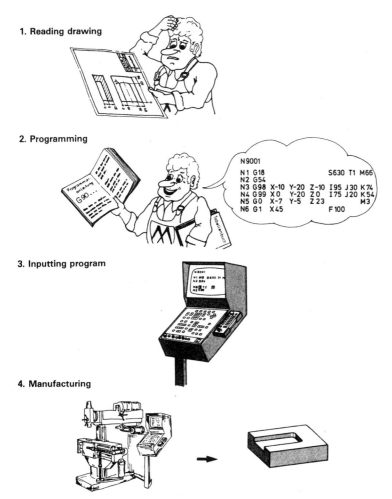

1. Reading drawing

2. Programming

3. Inputting program

4. Manufacturing

Fig. 1-2-14 The major steps involved in planning and machining a part.
(Deckel–Maho, Inc.)

4. **Producing the Part (Manufacturing)**
— **In conventional machining,** the operator turns the machine slide control handles to produce the size and shape of the part.
— **In CNC machining,** the part program in the offline or machine tool computer control moves the machine slides automatically to produce the part to the accurate size and shape.

ADVANTAGES OF CNC

CNC has grown at an ever-increasing rate, and its use will continue to grow because of the many advantages that it offers industry. Some of the most important advantages of CNC are illustrated in Fig. 1-2-15.

1. *Greater Operator Safety*

CNC systems are generally operated from a console that is usually away from the machining area; therefore the operator is not exposed to moving machine parts or to the cutting tool.

2. *Greater Operator Efficiency*

Since a CNC machine does not require much attention, the operator can perform other jobs while the machine is running.

3. *Reduction of Scrap*

Because of the high degree of accuracy of CNC systems and the elimination of most of the human errors, scrap has been drastically reduced.

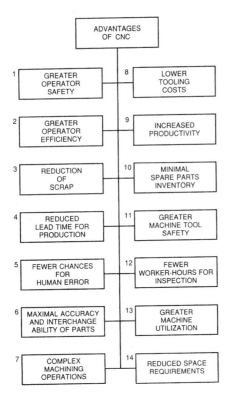

Fig. 1-2-15 Advantages of CNC in manufacturing.
(*Kelmar Associates*)

4. *Reduced Lead Time for Production*

The program preparation and setup for numerically controlled machines is usually short. Many jigs and fixtures formerly required are not necessary. The program can be stored in the computer memory or on removal media (disks, CDs, etc.) and used again as required for future production runs.

5. *Reduction of Human Error*

The CNC program eliminates the need for an operator to take manual trial cuts, make trial measurements, or make table positioning movements. There is no longer a need for the operator to change cutting tools, select the sequence of operations, and perform other routine functions.

6. *High Degree of Accuracy*

CNC ensures that all parts produced will be accurate and of uniform quality. The improvement

of accuracy of the parts produced by CNC assures the interchangeability of parts.

7. *Complex Machining Operations*

Complex operations can be done more quickly and accurately with CNC and electronic measuring equipment.

8. *Lower Tooling Costs*

CNC generally requires simple holding fixtures and fewer tools; therefore, the cost of fixture design and manufacture may be reduced by as much as 70 percent.

9. *Increased Productivity*

Because CNC controls all the machine functions, parts are produced faster, with less setup and lead time.

10. *Reduced Parts Inventory*

A large inventory of spare parts is no longer necessary, since additional parts can be made to the same accuracy when the same computer program is used.

11. *Greater Machine Tool Safety*

The damage to machine tools as a result of operator error is virtually eliminated because there is less need for operator intervention.

12. *Less Inspection Required*

Because CNC produces parts of uniform quality, less inspection time is required. Once the first part has passed inspection, very little additional inspection is required.

13. *Greater Machine Use*

Because there less time is required for setup and operator adjustments, production rates can increase up to 80 percent.

14. *Reduced Space Requirements*

CNC requires fewer jigs and fixtures and therefore less storage space.

SUMMARY

- The use of CNC on machine tools has increased the actual time spent on removing material from 20% with conventional machines to 80% or higher.
- CNC turning centers operate mainly on two axes, the X and Z.
  - the X axis controls the cross motion of the cutting tool.
  - The Z axis controls the carriage or turret travel along the workpiece.
- CNC machining centers operate mainly on three axes.
  - The X axis controls the table movement left and right.
  - The Y axis controls the table movement towards or away from the column.
  - The Z axis controls the vertical movement of the spindle.
- In the incremental programming system (G91), all distances are given from the last or preceding point.
- In the absolute programming system (G90), all distances are taken from a starting or reference point.
- The main advantages of CNC are:
  - better part accuracy
  - increased productivity
  - reduced manufacturing costs
  - better machine tool use
  - reduced scrap

KNOWLEDGE REVIEW

Machines Using CNC

1. How has the addition of electronic controls affected the amount of time spent on removing metal?
2. Name four of the most common machine tools that use CNC.
3. What machine tool is programmed on only two axes?

Programming Systems

4. Compare the absolute and incremental systems with regard to positioning from point to point.
5. Define the zero, or origin point.
6. What do the X+, X–, Y+, Y– commands represent in: (a) Absolute mode (b) Incremental mode

Positioning Systems

7. Name two categories of positioning (control) systems.

8. For what purpose are the following used:
   (a) point-to-point
   (b) straight line control
   (c) 3D contouring

From Art to Part

9. How is the toolpath controlled in:
   (a) conventional machining?
   (b) CNC machining?
10. List the four major steps involved in producing a part.
11. What system should be used when it is necessary to produce many identical parts?

Advantages of CNC

12. List four of the most important advantages of CNC for each of the following:
    (a) Production rates
    (b) Part accuracy
    (c) Cost reduction

SECTION 2

Computers in Manufacturing

COMPUTERS IN MANUFACTURING

Throughout history, the development of new tools and machines helped to change the lifestyles of people and improve the standard of living throughout the world. Humanity has progressed through the Stone Age, Bronze Age, Iron Age, and Machine Age, and each period has helped to improve our lot in life and make it more productive. We are now in the Computer Age, and nothing in past history can compare to the effect that the computer is having on the way we live, work, and play, Fig. 2-1. The computer has made us more productive, provided us with more leisure time, removed the drudgery of home chores, and in general radically changed the way we live.

The introduction of computers into manufacturing has reduced the need to rely solely on the operator's skill in manufacturing processes and replaced it with high-precision, computer-controlled machinery, Fig. 2-2. While this helps to reduce human error, it does not reduce the production knowledge required of the machine tool operator. The CNC machine tool operator still has to understand the fundamentals of cutting tools and manufacturing processes as much as ever. But now there is a need to develop other skills in order to produce the most accurate parts in the least amount of time possible, with computer-assisted manufacturing techniques and hardware.

Fig. 2-1

Fig. 2-2

Unit 3

Introduction to Computers

The computer, Fig. 2-3-1, is an electronic tool that can handle data with amazing speed, accuracy, and reliability. It does not have a brain and therefore cannot think for itself. The computer is only an *extension* of a person's brain and must be told exactly, in language that it understands, what it is to do. The computer can be likened to a machine tool, which is an extension of a person's muscles. The computer is fast, accurate, and stupid, where a human is slow, inefficient, and brilliant. A very useful manufacturing tool has evolved by combining the brilliance of a human with the speed and accuracy of a computer. Today's computers have become part of our everyday life and we would have great difficulty surviving without them.

In manufacturing, the computer has become a valuable tool that is very important to industries if they expect to compete in the marketplace locally or globally. It has made it possible to consistently manufacture high quality products at speeds undreamed of in the past.

OBJECTIVES

After completing this unit you should be able to:

1. Name and state the purpose of five main parts of a computer system.
2. Explain why a computer is so valuable to manufacturing industries.
3. Understand why both hardware and software are required for a computer system to operate.
4. Name the two most common computers used in industry.

Fig. 2-3-1 Computers, plus the ability of humans to reason, can combine to make a useful manufacturing tool.

KEY TERMS

central processing unit   memory storage
controller   output device
hardware   processor unit
input devices   software

THE EVOLUTION OF COMPUTERS

Computers are general-purpose machines that process data according to a set of instructions stored internally either permanently or temporarily. Computers and all the physical things attached to it are called **hardware**, and the instructions that tell it what to do are called **software**. A set of instructions that perform a particular operation is called a **program or software program**. The computer's ability to call in instructions and follow them is known as the **stored program concept.**

Since the computer's inception in 1947, it has progressed through various stages from a huge machine to small pocket-size versions that perform thousands of functions, in a span of about 50 years. A brief history of the development of the computer follows:

First-Generation computers, beginning with the UNIVAC 1 in 1951, were very large units that used vacuum tubes, and their memories were made of thin tubes of liquid mercury and magnetic drums. They were huge machines filling an entire room that had to be climate controlled, Fig. 2-3-2.

Second generation computers, in the late 1950s, replaced tubes with transistors and used magnetic cores for memories (IBM 1401, Honeywell 800). Their size was greatly reduced and with improvements they became much more reliable than the earlier models.

Third generation computers, beginning in the mid-1960s, used the first integrated circuits (IBM 360, CDC 6400) and the first operating systems and DBMS. Online systems were widely developed although most processing was batch oriented using punched cards and magnetic tapes.

Fourth generation computers, in the mid-1970s, used microchips almost exclusively and were often called **electronic brains**. This development made the microprocessor and the personal computer possible, Fig. 2-3-3, and introduced distributed processing, office automation, Query language, report writers, and spreadsheets that put large numbers of people in touch with computers for the first time.

Fig. 2-3-2 The first computer for commercial use was the IBM UNIVAC 1. *(IBM Corporation).*

Fig. 2-3-3 Thousands of transistors and circuits can be included on a tiny silicon chip. *(Rockwell International)*

Fig. 2-3-4 The lap–top and pocket computer have been made possible through many years of development.

Fifth generation computers, in the mid-1990s, started to use voice recognition, natural and foreign language translation, optical disks, and fiber-optic networks. The lap-top and pocket computers of today, Fig. 2-3-4, present a sharp contrast to the size, speed, and capabilities of the early computers. It is predicted that smaller and higher-speed computers, combined with better software, will allow the average computer to talk to us with reasonable intelligence early in the 21st century.

THE ROLE OF THE COMPUTER

The first computer, developed in 1947, was large and subject to breakdowns due to the many vacuum tubes it contained. As the electronics industry progressed from vacuum tubes to transistors, solid-state components, integrated circuits (ICs), and then microprocessors, computers became smaller, more reliable, and less expensive. Today computers no larger than a typewriter, and even lap-top and pocket-size models, are commonplace in industry and in many homes.

Some computers can perform millions of calculations per second because of the thousands of transistors and circuits jammed into the tiny chips (ICs), Fig. 2-3-3. Computer scien-tists can foresee the day when 1 billion transistors or electronic switches (with the necessary connections) will be crowded into a single chip. A single chip will have a memory large enough to store the text of 200 long novels. Advances of this type will decrease the capacity of computers considerably.

The Computer Microchip

Microcomputers are often referred to as "intelligent devices" or "electronic brains" because they sometimes function in a manner somewhat similar to that of the human brain. Every microcomputer has the ability to calculate and remember, just as humans do. But the microcomputer actually surpasses a human's abilities in that it can add single-digit numbers at a rate of 100,000 per second. It can memorize data precisely and it never forgets. A human's memory capacity is virtually limitless, but the microcomputer's memory is always limited by its very structure. A person can respond to a command or react to a situation, and make a decision on how to act. Of course, the microcomputer cannot do these things. Humans can see, hear, talk, and act, using their body to place them in touch with the outside world. The microcomputer, on the other hand, lacks the means to fully control its own input and output. It has only an interface which must be connected to man-made input and output equipment (keyboard, display screen, etc.). Thus, a microcomputer is inferior to humans in memory capacity, decision-making ability, and the ability to control input and output. But microcomputers have contributed enormously to numerous fields because of their ability to calculate and memorize information at amazing speeds.

Computing Speed and Memory

A microcomputer consists of a central processing unit **(CPU)**, a read only memory **(ROM)**, a random access memory **(RAM)**, and input/output ports **(I/O)** sections, Fig. 2-3-5. Using the numbers 2 and 6, let's see how each section of the microcomputer plays its role in adding the two together. To make the story simple, let's compare 2 and 6 to parts, and "+" to an in-

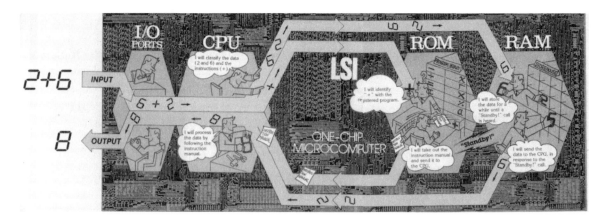

Fig. 2-3-5 The I/O receives and discharges information, the CPU processes and controls information, the ROM stores programmed information, and the RAM reads and writes data. *(Sharp Electronics, Inc.)*

struction. The microcomputer is a factory that processes the parts first by receiving the 2 and 6 through the I/O ports, the wicket. The CPU determines where to send and store each piece. The 2 and 6 are stored in RAM, the storehouse, while "+" is sent to ROM. ROM looks through the files, figures that "+" means "Add up!" and prepares an instruction manual. Now everything is ready to be put together in the CPU. The CPU starts processing the parts (returned by RAM) by referring to the manual given by ROM. This way 8, the end product, is forwarded through I/O.

Figure 2-3-5 shows how the information is processed within the computer. The system of the microcomputer is more complicated because a "=" must be given in order to get an answer. The microcomputer cannot receive data like 2 and 6 directly from a person but only through a computer keyboard. Nor can it send a result like 8 directly, it needs a display screen or equivalent. When adding or subtracting, the numbers are expressed in the decimal scale and the computer keyboard translates these into binary-scale electric pulses.

In the 1990s, some computer scientists felt that the prototype of a thinking computer incorporating *artificial intelligence* (AI) will be developed and become a commercial product about five years later, Fig. 2-3-6. This machine will be able to recognize natural speech and written language and will be able to translate and type documents automatically.

COMPUTER USE

Today computers are being used in a wide variety of activities that involve our everyday life, Fig. 2-3-7.

- **Business**—Department store computers list and total your purchases, at the same time keeping their inventory up to date and advising the company of people's buying habits.

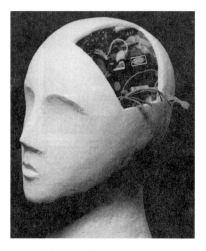

Fig. 2-3-6 Artificial Intelligence systems may soon be able to come close to the capabilities of the human brain. *(Society of Manufacturing Engineers)*

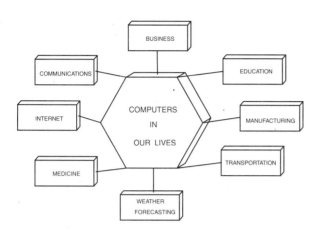

Fig. 2-3-7 The many ways that computers are affecting the lives of humans throughout the world. *(Kelmar Associates)*

Fig. 2-3-8 CAD systems are valuable to engineers and designers who research and develop new products.

— Computers have relieved accountants of the drudgery of repetitious jobs such as payroll processing.
— Many office workers today work from their home on a company computer and it is expected that this trend will increase in the future, eliminating the need to travel long distances to work and the need for baby-sitting services for instance.
• **Education**—Children are learning more at a younger age and future generations will have much broader and deeper knowledge than those of past generations.
   — School computers record students' records, courses, grades, and other information.
   — In the past, humanity doubled its knowledge every 25 years; with the computer the knowledge doubles every 18 months.
   — With this greater knowledge, the human race will explore and develop new sciences and areas that are unknown to us today.

Manufacturing

In the manufacturing industry the computer has contributed to the efficient manufacture of many goods. It appears that the impact of the computer will be even greater in the years to come. Computers will continue to improve productivity through *computer-aided design* (CAD), by which the design of a product can be researched, fully developed, and tested before production starts, Fig. 2-3-8. *Computer-assisted manufacturing* (CAM) results in less scrap and more control reliability through the computer control of the machining sequence and the cutting speeds and feeds.

Robots, which are computer-controlled, are being used by industry to an increasing extent. Robots can be programmed to paint cars, weld, feed forges, load and unload machinery, assemble electric motors, and perform dangerous and boring tasks formerly done by humans.

Virtual NC® is an interactive 3D simulation system that allows a person to visualize and analyze the operation of a machine tool, its CNC controller, and the material removal process, on a computer screen.

- **Medicine**—Medical centers store data on all known diseases with their symptoms, known cures, etc.
  - Doctors are able to patch their computer into the central computer and get an immediate and accurate diagnosis of the patient's problem, thus saving many hours and even days awaiting the results of routine tests.
- **Weather and Transportation**—The computer is used to forecast weather, guide and direct planes, spaceships, missiles, and military artillery, and monitor industrial environment.
  - In air defense control systems, the position and course of all planes from the network of radar stations are fed into the computer along with the speed and direction of each.
- **Internet and World Wide Web** - Computer networks have made possible the **Information Highway** that allows users to exchange mail and transfer information and electronic files to anyone in the world.
  - The World Wide Web has allowed users to visit any library, research center, and industries in the world to research the latest technological developments.

NC to CNC

An NC unit developed in 1968 required about 400 circuit boards loaded with transistors to operate, but a CNC unit of today only requires one circuit board to perform the same tasks.

An easy way to understand the difference between NC (numerical control) and CNC (computer numerical control) is as follows: NC + the Computer = CNC. Developed in the early 1970s, the growth of CNC has been very rapid throughout the world and has had a big effect on the design of machine tools. CNC has made it possible to control automatic loading and unloading of cutting tools, pallet changes, workstations, and inspection systems that can check part quickly and accurately. CNC has made highly-efficient manufacturing possible by improving the quality of manufactured parts, reducing scrap, improving productivity, and reducing manufacturing costs.

THE COMPUTER

Although it is not essential for a person to fully understand the workings of a computer, it is important to be familiar with the five main functional units of a computer system and their terminology.

Hardware and Software

There are two main components that make a computer do the things that they do; they are the hardware and software elements, Fig. 2-3-9.

1. **Hardware** - The main hardware of a computer is made up of microprocessors (16 to 128-bit technology), integrated circuits, and electronic memory modules. Very large-scale integrated circuits are available that are very compact, reliable, and are easy to maintain. Electronic interference can cause a computer to malfunction and most computers today are provided with housings that offer protection from electromagnetic and electrostatic interference. However, care should be used to keep abnormal electromagnetic and electrostatic interference as far as possible from the computer.

- The **hardware** consists of the mechanical and electronic parts, the power supply, and the peripheral equipment such as the reader, printer, and external memory units.

Fig. 2-3-9 **Hardware and software are the two main components that make it possible for a computer to operate.** *(Dell Computers)*

2. **Software** - The CNC software, also known as the system software, affects the intelligence or capabilities of the computer and coordinates all the operations performed by each hardware module or unit. It controls programming, graphic simulation of the cutting process, and the operation of the CNC machine tool. The CNC software makes all the hardware components perform as a computer.

3. **Electronic Memory Modules** - Electronic memory modules are used in computers to store information in two locations; a permanent storage for the CNC operating system that cannot be changed easily, and a temporary storage that can be easily retrieved and changed as necessary. Several memory types are used in computer systems, the most common being:

- **RAM** (Random Access Memory), also called direct access storage, is a type of memory storage unit into which data can be written (stored) and accessed (read).
- **ROM** (Read Only Memory) is a computer storage unit from which data (under normal conditions) can only be read. This type of data, usually installed by the manufacturer, contains the CNC operating system, plus canned cycles, or fixed routines, etc., that cannot be changed by the average person.

A computer system generally consists of three basic units that are important to its ability to process information, Fig. 2-3-10.

1. **Input Devices** consist of anything that is used to enter data or programs into a computer, including keyboards, disk drives (floppy and CD), tape-based devices, and punched-tape readers. They all are used to supply alphanumeric data to the computer in digitized form and at high speed.

2. **Controller** - The controller, memory storage, and processor units form the Central Processing Unit (CPU) which processes input data. The controller generally consists of a circuit board with a microprocessor chip, and several memory mod-

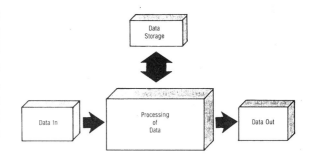

Fig. 2-3-10   The three basic functions of a computer are to accept data, process data, and output data.
(Kelmar Associates)

ules for data storage. The internal hard drive on a computer provides memory storage for the permanent and long-time storage data.

3. **Output Devices** are any devices that take computer information and change it into a format that is easily understood by anyone. The computer screen, printer, and plotter are the most common output devices.

FUNCTIONS OF A COMPUTER

Computers fill three major roles in CNC.

1. Most machine control units (MCUs) include a computer or incorporate a computer in their operation. These control units are generally called *computer numerical controls* (CNC).
2. Most of the part programming for CNC machine tools is done with off-line computer assistance.
3. There is an increasing number of machine tools that are controlled or supervised by other computers, which may be in a control room or even in another plant. This arrangement is more commonly known as *direct numerical control* (DNC).

The computer has also found many uses in the overall manufacturing process. It is used for such things as part design (CAD), testing, inspection, quality control, planning, inventory control, gathering of data, work scheduling, warehousing, and many other functions in manufacturing. The computer is having, and will continue to have, profound effects on manufacturing processes in the future.

TYPES OF COMPUTERS

Most computers fall into two basic types, analog or digital. The *analog* computer does not work directly with numbers and has been used primarily in scientific research and problem solving. Analog computers have been replaced in most cases by the digital computer.

Most computers used in industry, business, and in the home are of the electronic *digital* type. It is this type that is used in computer numerical control work and the one explained in this book. The digital computer accepts an input of information in numerical form, processes that information according to prestored or new instructions, and develops output data.

There are generally three categories of computers and computer systems: the mainframe computer, the minicomputer, and the microcomputer. Although each of these basically performs the same tasks, some are better suited to certain applications.

The *mainframe computer*, Fig. 2-3-11, which can be used to do more than one job at the same time, is large and has a huge capacity. It is generally a large company's main computer and the one that performs general-purpose data processing, such as CNC part programming, payroll, cost ac-

Fig. 2-3-12 The minicomputer is generally a dedicated type of computer that performs specific tasks. *(Bausch and Lomb)*

counting, inventory, and many other applications. The mainframe computer generally has a number of individual keyboard terminals connected to it, and each of these can feed information to the mainframe computer simultaneously. The CPU is designed to accept word lengths of at least 32 bits (a bit is a binary digit, either a 1 or a 0).

The *minicomputer*, Fig. 2-3-12, is generally smaller in size and capacity than the mainframe computer. This type of computer is generally of the "dedicated" type, which means that it will perform a specific task such as the:

1. Generation of part programs
2. Distribution of program data for a part of various CNC machines (DNC)
3. CNC of a single machine tool (CNC)
4. Management of inventory and scheduling

The minicomputer's CPU is generally designed to accept a maximum word length of 32 bits; however, some of the more recent models can accept up to 64 bits.

The *microcomputer* or personal computer, Fig. 2-3-13, generally contains one chip (a microprocessor) that contains at least the arithmetic-logic and the control-logic functions of the CPU. The microprocessor is generally designed for simple applications and must be accompanied by other electronic devices (usually on a printed circuit board) for more complex applications. The CPU is usually designed to accept 16 bits; however, today there are 32-bit microprocessors available.

Fig. 2-3-11 The mainframe computer generally has a large capacity and is usually a company's main computer. *(Hewlett Packard)*

Fig. 2-3-13 The microcomputer's processing unit consists of one or more microprocessors, storage, and input and output facilities. *(Society of Manufacturing Engineers)*

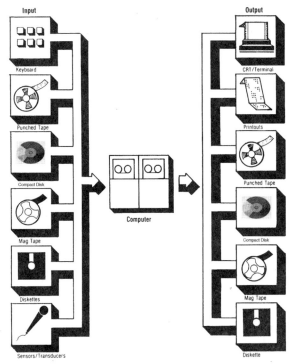

Fig. 2-3-14 Types of media and equipment used to input and output information to and from the computer. *(Modern Machine Shop)*

Computer Functions

The function of a computer is to receive coded instructions (input data) in numerical form, process this information, and produce output data which causes a machine tool to function. Many methods are used to put information into a computer. Some of the more common are perforated tape, magnetic tape, floppy disks, compact disks (CDs), digital versatile disk (DVDs) and specially designed sensors, Fig. 2-3-14. At this time, the most commonly used method to input data used in CNC is the computer keyboard.

Output data (information coming out of the computer) may be saved as punched tape, printout sheets, magnetic tape, or transferred directly to another computer. Large amounts of information are generally stored on floppy, digital tape storage devices, or compact disks. Output data may also be sent directly to the MCU and then to the servomechanisms that activate the machine tool slides.

SUMMARY

- The computer is an amazing tool that can process large amounts of information and make complicated calculations at amazing speeds.
- The computer is widely used in almost every-thing that affects our everyday life including business, education, medicine, manufacturing, transportation, weather forecasting, etc.
- NC became CNC with the development of the computer and this provided industry with a valuable tool that increased product quality, productivity, and reduced manufacturing costs.
- Input devices are used to enter information (data or programs) into a computer, the controller processes the information, and the output devices change the information into a format that is easily understood.
- The digital computer is most commonly used and is available in three categories: mainframe, minicomputer, and microcomputer.
- The main purpose of a computer is to receive coded instructions, process this information, and produce data that can be easily understood.

KNOWLEDGE REVIEW

Evolution of Computers

1. Define the terms: hardware, software, program.
2. What is the name of the first computer developed?
3. What effect did the microchip have on computer development?

Role of Computer

4. How is artificial intelligence (AI) expected to affect the computer of the future?
5. List the four main sections of a microcomputer.
6. Briefly describe how a computer processes numbers.

Computer Use

7. List four areas where computers have affected our personal life.

NC to CNC

8. How did NC become CNC?
9. Name four things that CNC can control on a machine tool.

The Computer

10. State the main purpose of the following:
    (a) input devices
    (b) controller
    (c) output devices
11. What are the three main hardware components of a computer?
12. List three operations that computer software controls.
13. Define the RAM and ROM acronyms and briefly state the purpose of each.

Types of Computers

14. Where are analog and digital computers used?
15. Name three categories of computers and computer systems.

Unit 4
Input Media

Computer Numerical Control (CNC) operates machine tools by sending a series of coded instructions consisting of alphabet letters, numbers, and symbols in language that the machine control unit (MCU) can understand, Fig. 2-4-1. These coded instructions are turned into pulses of electrical current or other output signals that operate the motors and servomechanisms of the machine tool. The coded instructions or commands are listed in a logical sequence to cause a machine tool to perform a specific task or a series of tasks in order to produce a finished product.

OBJECTIVES

After completing this Unit you should be able to:

1. List three commonly used CNC input media.
2. List two advantages of using diskettes as the input media.
3. Know the sequence of operations to follow in order to produce a CNC program.
4. Understand how CAD/CAM software is used in designing a part and creating the CNC machining codes.

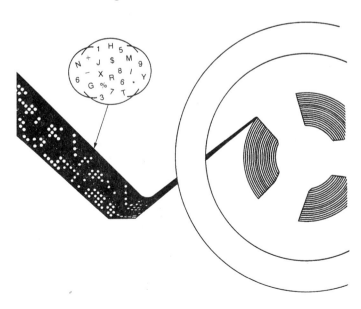

Fig. 2-4-1 CNC sends a series of coded instructions (alphabet, numbers, and symbols) to the MCU to control a machine tool.
(Kelmar Associates)

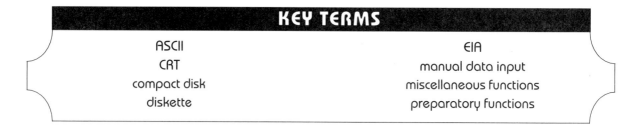

TYPES OF INPUT MEDIA

As CNC developed over the years, numbers of different media were used to convey the information from a drawing to the MCU (machine control unit). Early types of input media were manual, punched card, magnetic tape, punched tape, Fig. 2-4-2. Punched card and magnetic tape are rarely used today, but punched tape may still be used in older NC and CNC machines. The most common input media used for most CNC machines is manual data input, floppy disks, and compact disks.

Manual Data Input (MDI). Data that is input through the computer keyboard can be used to program the control system by setting the dials, switches, push buttons, etc. Although this

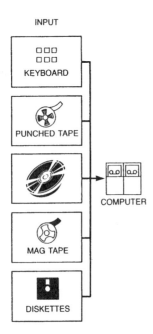

Fig. 2-4-2 Types of input media that have been used to transfer data into the MCU *(Modern Machine Shop)*

method is not often used because it is slow and subject to operator error, most CNC machine tools can be programmed manually, especially for setup purposes.

Punched Cards. In the early days of NC, some systems used 80- or 90-column punched cards that carried a great deal of information. The disadvantage of punched cards was that the cards were bent easily, would sometimes jam in the card reader, and the sequence of the cards could be altered during handling.

Magnetic Tape. In the late 1950s, some NC machine control units (MCUs) used 1 in. (25 mm) magnetic tape to store data. This tape was similar to what was used to record music and conversation, but was of a higher quality. It was not used for any length of time because interference from nearby electrical equipment such as transformers and other shop equipment had a tendency to erase or scramble some of the information on the tape. Today magnetic tape has better shielding from outside electrical interference, and the ¼ in. (6 mm) tape in a cassette is used for some CNC applications.

Punched Tape. At one time punched tape was the most common input medium and the standard for the NC industry. The Electronic Industries Association (EIA) selected the 1 inch (25 mm) eight-channel tape, Fig. 2-4-3, using the binary-coded decimal (BCD) system as the standard. Holes punched in the channels and rows determined the operating instructions for the machine tool. A tape reader on the MCU (machine control unit) decoded the hole patterns and converted them into electrical pulses that operated the motors and the servomechanisms of the machine tool.

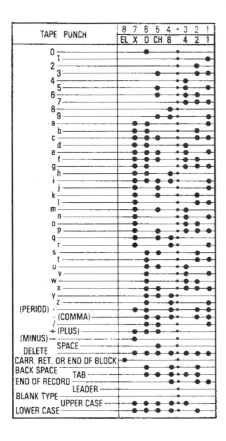

Fig. 2-4-3 EIA standard coding for 1-inch-wide eight-channel tape.

Punched tapes are made from several types of materials, including paper, Mylar (a type of plastic), and aluminum foil.

1. **Paper**: Early tapes were made of paper, however, because they required care to prevent them from being torn, they were generally only used for filing purposes.
2. **Mylar:** Tapes made of Mylar, laminated between two strips of paper proved sturdy and almost impossible to tear and they quickly became popular. Although more expensive than paper tape, Mylar was highly recommended for shop use because of its strength.
3. **Foil:** Aluminum foil tapes are sturdy but not generally recommended because they are very hard on the punches of the tape-preparation equipment.

The storage media available today are compact, reliable, and have far greater storage capacity than the punched tape and floppy disk of not too many years ago. Some of the more common media are shown in Fig. 2-4-4 and each is briefly explained to provide the reader a working knowledge of each type.

1.44 Drive - uses a reusable magnetic storage disk enclosed in a sturdy plastic jacket, and contains channels or tracks on which magnetic bits of information can be stored. A reading head on the MCU can decode the bits of information on the diskette. The floppy disk (diskette) used today, is a rigid 3.5 in. microfloppy that holds up to 1.44 MB (megabytes or one million bytes) of information, Fig. 2-4-4A.

Zip Drive - a 3.5 in. removable and reusable disk that can hold up to 100 MB of data, Fig. 2-4-4B. The drive is included with software that can catalog the disks and lock the files for security. A new 250 MB version of the Zip drive, introduced in 1998, can also read and write the 100 MB Zip cartridges.

Super Disk Drive - similar to the Zip drive but can handle both types of floppy disks and has a capacity of up to 120 MB of memory, Fig 2-4-4C.

Syquest - a drive for removable disks that has a capacity of up to 1 Gb (gigabytes or one billion bytes) pieces of data. This company sold its assets to Iomega in early 1999, Fig. 2-4-4D.

CD-ROM (Compact Disc Read Only Memory) - a compact disk format, have digital data carved into them with a laser. Standard CDs can very quickly record or retrieve up to 650 MB of data, or about 250,000 pages of text, Fig. 2-4-4E.

DVD (Digital Versatile Disk) - a family of double-sided optical disks that have as much as 750% more capacity than a CD. DVD is like a large CD-ROM that can handle data, audio, and video formats faster and with improved clarity, Fig. 2-4-4E.

Ethernet - is a high-speed data link protocol that uses CSMACD (Carrier Sense, Multiple Access, Collision Detect) technology to transmit data to connecting stations. It can transmit data at speeds of 10 million bits per second over heavy coax cable or fiber optic cable. Ethernet is the most widely used LAN (Local Area Network) system with unlimited capacity that provides the user with a secure, private 10-megabit data path, Fig. 2-4-4F.

RS 422 - are standard interfaces that provide multipoint connections for connecting serial devices used to replace older RS 232 interfaces. They support higher data transfer rates and have greater resistance to electrical interference, Fig. 2-4-4G.

TAPE-CODING SYSTEMS

Two CNC tape-coding systems have been developed over the years and at one time were considered industry standards. These are the Electronic Industries Association (EIA) system and the American Standard Code for Information Interchange (ASCII) system which are very similar. They both use the BCD system for numerical data and both use 1 inch (25 mm) eight-channel tape. There is a demand for both types of coding, and

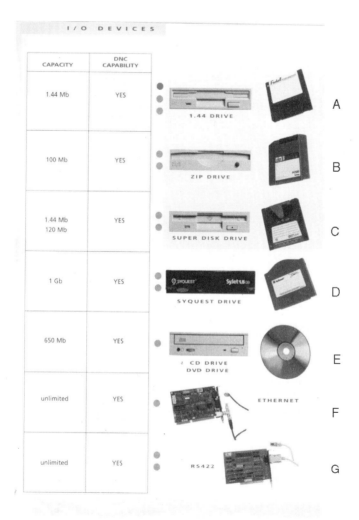

CAPACITY	DNC CAPABILITY		
1.44 Mb	YES	1.44 DRIVE	A
100 Mb	YES	ZIP DRIVE	B
1.44 Mb 120 Mb	YES	SUPER DISK DRIVE	C
1 Gb	YES	SYQUEST DRIVE	D
650 Mb	YES	CD DRIVE DVD DRIVE	E
unlimited	YES	ETHERNET	F
unlimited	YES	RS422	G

I / O DEVICES

Fig. 2-4-4 The drives and input media listed are the most common methods of storing, or entering information for a computer. *(Fadal Machining Centers)*

this does not present a problem since today's MCUs recognize the type of code and switch automatically. A comparison of the two coding systems appears in Fig. 2-4-5.

EIA Coding System

The EIA coding system for NC tape was used to control NC machine tools and equipment. It used a BCD system on 1 inch (25 mm) eight-channel tape for numerical data.

1. **Five of the channels (1 to 4)** are assigned values of 0, 1, 2, 4, and 8, so that any value from 0 to 9 can be inserted in one line of the tape.
2. **The fifth channel** (indicated by CH) is the odd-parity bit and is used as a safety device to reduce the chance of error when preparing tape.

3. **The sixth channel** (indicated by 0) is a special binary code and is assigned to lowercase letters of the alphabet or when a 0 is required.
4. **The seventh channel** (X) indicates the minus sign (–) and is also assigned to lowercase letters of the alphabet.
5. **The eighth channel** (EL) indicates the end of a block of information. It is punched at the beginning and end of most NC punched tapes and is never combined with any other holes.

ASCII Coding System

Various committees working with the American National Standards Institute developed the ASCII system. The goal was to produce a coding system that could be used by telephone and telegraph companies, computer companies, governments, and others as an international standard for all information processing and communications systems. This collaboration has resulted in making the ASCII system the **universal** perforated tape coding system. However, NC has been and still is heavily committed to the original EIA (RS-244) code. The ASCII subset used for NC tape is now the EIA (RS-358) Recommended Standard. See Fig. 2-4-5 for a comparison of the EIA and ASCII tape-coding formats.

Since both tape-coding systems are used and the equipment is capable of handling either, it is important to understand the differences found in each system.

Similarities

- EIA and ASCII both use the 1 inch (25 mm) eight-channel tape.
- The codes for digits 0 to 9 are the same for both systems.
- Both systems use the binary coded decimal (BCD) system for numerical data.

Differences

- ASCII provides coding for both uppercase and lowercase letters of the alphabet. EIA provides coding only for lowercase letters.

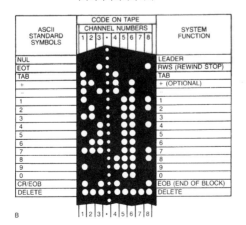

Fig. 2-4-5 A comparison of the (A) EIA and (B) ASCII coding systems for NC tape. (*The Superior Electric Co.*)

- ASCII uses an even-parity check (channel 8), so there must always be an even number of holes in each row. EIA uses an odd-parity check.
- In EIA, holes are punched in two additional channels (5 and 6) to identify the numbers and certain symbols.
- In EIA, a hole is punched in channel 7 for all alpha characters such as the alphabet.
- Parity check is used to confirm that holes have been punched correctly.

PROGRAM PREPARATION

A CNC system is an obedient servant that will faithfully follow the instructions it receives, but it cannot think for itself and cannot overcome errors in programming. CNCs may use either absolute or incremental positioning; with absolute positioning, each machining location is given in relation to a zero point, or origin. When the incremental positioning is being programmed, each machining location is given from the last position.

The stages in preparing a CNC program begin with the part drawing, and progress through preparing a manuscript, writing the program, and checking the accuracy of the program to make it ready for use in a CNC machine, Fig. 2-4-6.

Part Drawing

The technical or engineering part drawing must contain all the specifications of the part, such as

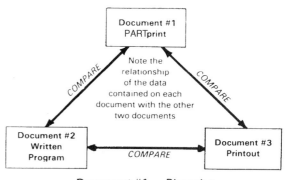

Document #1 — Blueprint

Fig. 2-4-6 The three CNC documents that contain information about the part and its machining requirements. *(The Superior Electric Company)*

the shape, sizes, dimensions, tolerances, and the type of operations that must be performed. In Fig. 2-4-7, six holes of .375 inch diameter must be drilled in a 3.625 × 6.000 inch plate. It is assumed that the plate has previously been machined to size; only the holes must be located in the proper location and drilled.

Manuscript

A CNC programmer must take the information contained on the part drawing (print) and put it into a language that the MCU will understand. This is done by taking the information from the part drawing and preparing a program manu-

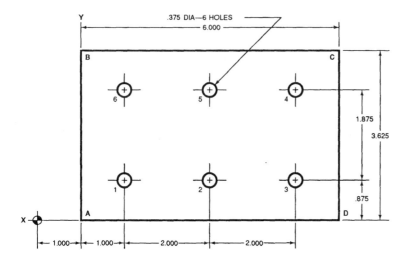

Fig. 2-4-7 The part drawing should contain all the information required to produce the part to the exact sizes. *(Kelmar Associates)*

CNC Program Worksheet

Part # *CNC-133A*
Program # *06789*

N	G	X (J)(D)	Y (K)(S)	Z	F (L)(T)(H)	M	Remarks
010	20	G90					
020	00	G-54 X·500	Y/·560			M03	
030	43			Z·/00	H0I	M08	
040	0I		S2000	Z-·/25	F5·0		
050			Y·500				
060		X/·500					
070			Y/·000				
080		X/·060					
090	00			Z·/00		M09	
/00	28			Z·/00		M05	
/10	28	X/·060	Y/·060				
/20						M30	
%							REWIND CODE

Fig. 2-4-8 A manuscript should list the sequence of operations, tools required, and the various sizes of a part. (*Kelmar Associates*)

script that lists the steps and sequences required to machine the part, Fig. 2-4-8.

The programmer plays a very important part in whether a workpiece will be produced to exact sizes. Since a CNC machine does exactly what it is programmed to do, it must be programmed correctly.

Before a program is prepared, it is necessary to know whether the program is being prepared for absolute or incremental positioning. Since the role of the programmer is so essential for the successful operation of any CNC machine tool, it is desirable to review the qualities a person should have to become a successful programmer, Fig. 2-4-9.

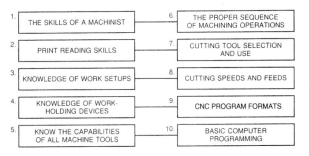

1. THE SKILLS OF A MACHINIST
2. PRINT READING SKILLS
3. KNOWLEDGE OF WORK SETUPS
4. KNOWLEDGE OF WORK-HOLDING DEVICES
5. KNOW THE CAPABILITIES OF ALL MACHINE TOOLS
6. THE PROPER SEQUENCE OF MACHINING OPERATIONS
7. CUTTING TOOL SELECTION AND USE
8. CUTTING SPEEDS AND FEEDS
9. CNC PROGRAM FORMATS
10. BASIC COMPUTER PROGRAMMING

Fig. 2-4-9 A list of desirable qualities of a good programmer. (*Kelmar Associates*)

In many cases, it is advisable to make a programmer's sketch of the part showing the sequence of operations and dimensioning it to suit the programming mode. The following steps are required to produce a manuscript (program).

1. Select the fixture most suitable to locate and hold the workpiece on the machine table.
2. Select the zero or origin, to allow the workpiece and machine tool to be aligned.
— This point is usually at the bottom left corner of the part, the center of a round part, or any point on the holding fixture, or any point off the job.
3. Select the first tool change point where the cutting tools can be changed and workpieces loaded and unloaded.
— This location is usually a second point away from the part where tools can be loaded or unloaded easily and without interference.
4. Determine the sequence of machining operations necessary to machine the part accurately and in the least amount of time.
5. Record the sequence of operations on the manuscript (program) form.
— Enter operator's instructions such as tool changes or speeds.

For an example of the steps a programmer must follow, see Fig. 2-4-7. In this example, point-to-point positioning in the absolute mode and the interchangeable (compatible) format will be used to prepare the program to drill the six holes. For simplicity in this manuscript, let us assume that the holes will not be center-drilled before the .375 inch holes are drilled, and no tool changes are required.

Each horizontal line on the manuscript, Fig. 2-4-10, contains one block of information for one positioning movement along with the miscellaneous functions (tool changes, speeds, coolant, etc.) that are required. Let us examine each block of information.

N010-The Sequence Number. Each block of information is assigned a sequence number, and is the first item entered on the manuscript.

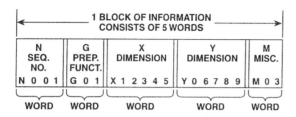

Fig. 2-4-10 A block of information should contain information for one positioning movement along with the necessary miscellaneous functions.
(*The Superior Electric Company*)

— The sequence number identifies each block of information and allows the information to be recalled at any time for reference or revision.
— The sequence numbers are generally listed in progressive order by 10s, and can be preceded by the letters H, N, or O.

G81-Preparatory Function. The G code refers to functions such as positioning systems, the direction of spindle rotation, canned and fixed cycles, and programming mode. In this example, G81 refers to a drilling operation that is one of the fixed cycles.

Numbering Operations. Line numbers such as 010, 020, 030, etc., allow new information to be inserted between any operation without the need to renumber all of them.

X2.0 The amount of movement along the X axis is indicated in this location. In absolute positioning, all distance (movements) are taken from the zero or origin.

— A movement to the right of the zero point is a plus movement; movement to the left is a minus movement.
— Plus movements do not have to be entered with the plus symbol (+) but minus movements must be indicated with the minus symbol (-) on the manuscript.
— In this example, the cutting tool must be moved 2.000 inch to the right of the zero point.

Y0.875 The amount of movement along the Y axis is indicated in this location.

— A movement up or away from the zero point would be a plus movement.
— A movement down or toward the zero point would be a minus movement.
— In this example, the cutting tool must be moved .875 inch up from the zero point.

Z1.0 The amount of movement along the Z axis either toward or away from the Z zero or gage height would be a plus or minus movement.

— A plus movement would be above the Z zero, and a minus movement would be below the Z zero.
— In this example, the spindle of the machine holding the drill would move 1.000 inch above the Z zero.

M03-Miscellaneous Function. Any miscellaneous function codes such as tool change, tool description and size, rapid feed rate, rewind codes, etc., are indicated in this location. In this example, the M03 code will cause the cutting tool to revolve in a clockwise direction.

Remarks- Operator Instructions. This column is used to inform the machine tool operator of tool changes, types, sizes of cutting tools, or other actions that are necessary.

PREPARING THE PROGRAM

After the manuscript has been completed and checked for accuracy, the program can be prepared in a number of ways.

Manual Preparation

In the early days of NC, perforated (punched) tape was produced on tape-preparation equipment similar to a typewriter. The programming assistant would key in the information from a manuscript on the tape typewriter, which would punch holes in the proper location on the tape. All the data required to produce a part was entered and then it was checked for accuracy.

Although some of the older tape typewriters may still be in use, most programs are prepared on a computer by keying in the information from

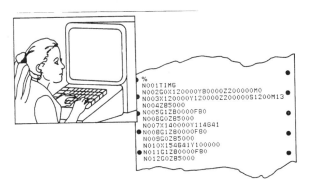

```
%
N001TIMG
N002G0X120000Y80000ZZ00000M0
N003X120000Y120000ZZ00000S1200M13
N004Z85000
N005G1ZB0000FB0
N006G0ZB5000
N007X140000Y114641
N008G1ZB0000FB0
N009G0ZB5000
N010X154641Y100000
N011G1ZB0000FB0
N012G0ZB5000
```

Fig. 2-4-11 **Information on the manuscript is keyed into a computer to produce the program that will machine the part.** *(Modern Machine Shop)*

a manuscript. The information is keyed in from the manuscript starting at the top of the sheet, Fig. 2-4-11.

— Each line of the manuscript represents one block of information on the tape and in the computer program.

— Once the program has been entered, it is checked and corrected if necessary, and then the computer activates the tape-punching unit to produce the tape.

— While the program is in the computer, it can also activate the printer to produce a printout record of the program.

Computer Preparation

CNC programs are generally prepared by computer systems with very little help from the programmer. For example, a circuit board containing 2000 to 3000 holes can be designed using computer-aided design (CAD) equipment. This equipment can produce both the technical prints (drawings) and artwork necessary for the manufacture of circuit boards and the program to run the drilling machine that produces the holes. With this system, no manual programming is required, because the location for each hole is known from the generated artwork. From the data, the computer prepares a program similar to the one produced by a programming clerk on tape preparation equipment.

CAD/CAM Software

The introduction of the computer and CAD (computer-aided design) software programs made it possible for designers and engineers to create a part's drawing on the computer screen. CAM (computer-aided manufacturing) software can use the information from the CAD database to create the machining codes, assign cutter toolpaths, and machine the part. See Units 15 and 19 for examples of CAD/CAM applications for the CNC Mill and Lathe.

Editing Programs

Regardless of how a program has been prepared (manually or computer assisted), it must be checked for accuracy and be available for engineering changes. It is quite possible that a new program could have some errors (bugs) in it, and it is wise to discover these before the work is scrapped or the machine damaged.

The programmer must review the manuscript to check for errors, wasted machine motion, and repetition of unnecessary information.

Wherever possible, the manuscript and the computer printout should be compared to catch any obvious errors, Fig. 2-4-12.

The printout should also be used to check the accuracy of data transmitted over telephone lines to the CNC machine tool at another location.

Fig. 2-4-12 **The manuscript, printout, punched tape, and transmitted data must be checked for accuracy before it is used on a machine.** *(Hewlett Packard)*

Regardless of how carefully all sources are checked, some error can or may always slip through because of the human factor. There are several editing systems available that on the market which have been specifically developed to check the accuracy of a program before it is used on a machine.

1. CRT (Cathode-Ray Tube) Screens or Plotting Units

Today most of the programming systems use a minicomputer or microcomputer with either a CRT screen or a plotting device, Fig. 2-4-13. Visual images of the part can be created that will quickly show errors such as incorrect contours or a misplaced decimal in a dimension. A dimension of 2.5 in. could easily be either .250 or 25 inches, depending on the placement of the decimal, and this difference would become very obvious when viewed on the screen. The plotter device can be used to generate the cutter path to see that it produces the correct shape of the part.

2. Shop-Floor Programming Systems

Viewing screens and keyboards are now standard equipment for many computer numerical control (CNC) shop-floor programming units. The visual part can be seen on the screen at the machine tool, and any corrections or revisions can be made to the part before the machine tool is run.

Fig. 2-4-13 CRT screens and plotting devices can be used to check the accuracy of a CNC program. *(Hewlett Packard)*

3. Portable Editing Units

Portable editing units consisting of tape readers, punches, data readout, CRT screens, etc. have been developed. They can be taken directly to the machine tool where program changes may be necessary or where a program must be checked. This is especially useful when it is impossible or not convenient to take the program back to the programming area for corrections or revisions.

Today's programming systems and controls have made part programming, editing, and revising much easier than the early NC and CNC systems. Regardless of how good the equipment is, there is always the possibility of an error in the program because of its length and the human factor involved. Until a foolproof system is devised, some program editing and revising will always be necessary.

SUMMARY

- CNC operates machine tools by sending coded instructions to the MCU where they are turned into electrical impulses that operate the motors and servomechanisms.
- The most common input media is the computer keyboard, floppy, disks, and compact disks.
- The two common tape-coding systems, the EIA and ASCII, both use the 1 inch (25 mm) eight-channel tape.
- The main stages of preparing a CNC program consist of reading the part print, writing a manuscript, and writing the program.
- A block of information usually contains a sequence number, preparatory function, X, Y, and Z dimensions, and miscellaneous functions.
- CAD/CAM software makes it possible to design a part on the computer screen, create the machining codes, simulate the toolpaths for machining, and revise the program if necessary.
- CNC programs should be checked for accuracy before they are tried on a machine, to avoid errors and costly damage to the cutting tool or machine.

KNOWLEDGE REVIEW

1. Briefly explain how numerical control operates.

Types of Input Media

2. Name four early types of input media that have been used in NC.

3. Describe and state the advantages of CD and DVD disks.

4. Describe the standard input medium that has been selected by the EIA.

5. Which types of punched-tape material is almost impossible to tear?

6. What are the advantages of using diskettes for storing NC information?

Tape-Coding Systems

7. Name the two types of tape-coding systems used in industry.

8. What channels on the EIA tape are assigned to the numerical values of 1, 2, 4, and 8?

9. Compare the similarities and differences of EIA and ASCII numerical control tape.

Tape Preparation

10. Explain absolute and incremental positioning.

11. Name three documents that contain information regarding the part program.

12. What role does the part drawing play for CNC programming?

13. What is the purpose of a manuscript?

14. Why is the programmer so important to CNC work?

15. List five of the most important qualifications that a good programmer must have.

16. Briefly list the five steps required to produce a manuscript (program).

17. Briefly explain sequence number, preparatory function, and miscellaneous function.

18. Explain X and Y increments and the direction of movement for each when there is a plus or minus movement.

Preparing the Program

19. How can computers be used to produce a CNC program and greatly reduce the amount of time needed?

20. List three ways that CAD/CAM software assist programmers.

21. Why is it important that CNC programs be checked before they are used on a machine?

22. What should the programmer look for when editing a program?

23. How can a CRT screen or a plotting unit help to check the accuracy of a CNC program?

24. State the advantages of shop-floor programming systems and portable editing units.

SECTION 3

How CNC Operates Machine Tools

Unit 5 CNC Components

HOW CNC OPERATES MACHINE TOOLS

Computer Numerical Control (CNC) operates machine tools in the same way a skilled operator would manually, but it is done automatically through stored program data. The operator reads the part drawing and then operates the machine slide handles to machine a part to size and shape. The accuracy of the part produced depends entirely on the skill and experience of the operator. CNC overcomes the possibilities of human error because the machine's functions are controlled by a fixed program and are not dependent on operator skill.

CNC offers almost unbelievable savings in production costs and part accuracy, and many other benefits. The following steps summarize how CNC works:

1. Numerical data may be fed into the system by stored data through a floppy disk, compact disk, or directly from a computer.
2. A translating unit reads the data and changes it into an electrical form that the machine tool can understand.
3. A memory system stores the data until it is needed.
4. Rotary resolvers or servo units (transducers) on the machine tool convert the data into the required machine movements.
5. A gaging device measures the machine movements to determine if the servo units have given the correct commands.
6. A feedback unit feeds information back from the gaging device for comparison to see that the machine has moved the required amount.

Once the program for a particular part has been checked for accuracy, it is ready to use. The program contains, in binary form, complete information for moving the machine table to each machining location. No special jigs are required; a simple holding fixture is all that is needed to locate and hold the part. The computer program accurately guides the machine table to the correct location, stops there, starts the cutting tool, sets speeds and feeds, and performs all the necessary operations required to produce the finished part. CNC is used on all types of machine tools, electrical discharge machines, welding machines, and inspection systems, and for most manufacturing and assembly systems.

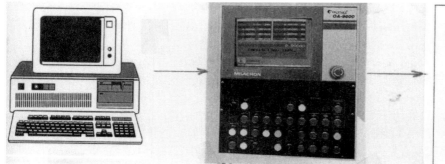

Table Top Computer   Control Unit

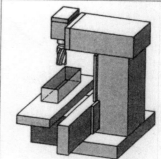

Simple Machine Tool
(e.g. Machine Model
for Training Purposes)

Unit 5
CNC Components

A CNC system consists of a number of components and the computer that carries out all the calculations and logical link-ups are at its heart. Since the CNC system is a linking element between the operator and the machine tool, the two interfaces are required for it to function properly, Fig. 3-5-1.

1. The **operator interface** that consists of the control panel and the various connections for the tape reader, floppy disk, compact disk, keyboard, and printer,
2. The **machine tool interface** that consists of a control interface, the axis control, and a power supply.

OBJECTIVES

After completing this unit you should be able to:

1. Name the five main components of a CNC system.
2. List the four benefits that have made CNC so widely used throughout the world.
3. List the advantages and disadvantages of open- and closed-loop systems.
4. Understand the functions of the machine control unit (MCU) and the central processing unit (CPU).

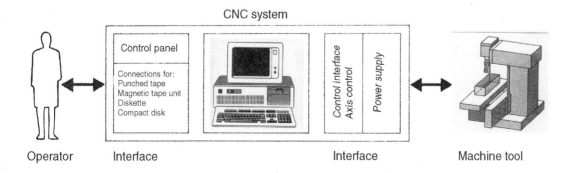

Fig. 3-5-1 The components of a CNC system that provide a means for the operator to communicate with the machine tool. *(Kelmar Associates)*

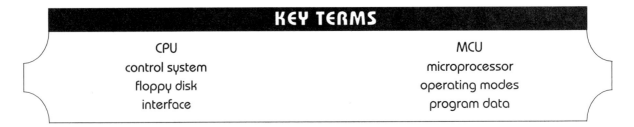

KEY TERMS

CPU	MCU
control system	microprocessor
floppy disk	operating modes
interface	program data

Starting with the edited program for a specific part, many elements are involved, from the control system to the machine tool, to decode and process this information and activate the machine tool that will produce a finished part. The key elements are the tape reader, NC control systems, servomechanisms, MCU, CPU, etc., Fig. 3-5-2. All these elements play a very important part in the NC of machine tools and each will be covered in detail later in the unit.

CNC PERFORMANCE

NC has made great advances since it was first introduced in the mid-1950s as a means of guiding machine tools through various motions automatically, without human assistance. The early machines were capable only of point-to-point positioning (straight-line motions), were very costly, and required highly skilled technicians and mathematicians to produce the tape programs. Great advances in CNC came as the result of new technology in the electronics industry. The development of transistors, solid-state circuitry, integrated circuits (ICs), and the computer chip have made it possible to program machine tools to perform tasks undreamed of as

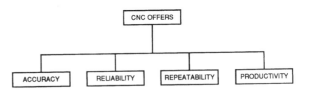

Fig. 3-5-3 CNC has been widely accepted because of the many advantages it has to offer industry. *(Kelmar Associates)*

recently as a decade ago. Not only have the machine tools and controls been dramatically improved, but the cost has been continually dropping. CNC machines are now within the financial reach of small manufacturing shops and educational institutions. Their wide acceptance throughout the world has been a result of their accuracy, reliability, repeatability, and productivity, Fig. 3-5-3.

Accuracy

CNC machine tools have been well accepted by industry because they are capable of machining to very close tolerances. At the time when NC was being developed, industry was looking for ways to improve production rates and achieve greater accuracy on products. A skilled machin-

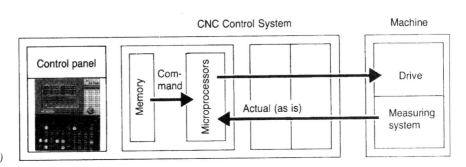

Fig. 3-5-2 The key elements of a CNC Control System. *(Kelmar Associates)*

ist is capable of working to close tolerances, such as .001 inch (0.025 mm) or even less on most machine tools. It has taken the machinist many years of experience to develop this skill, but this person may not be capable of working to this accuracy every time. Human error will always result in some mistakes, which means that the product may have to be scrapped.

Modern CNC machine tools are capable of producing workpieces that are accurate to within a tolerance of .0001 to .0002 inch (0.0025 to 0.0050 mm). The machine tools have been built better, and the electronic control systems ensure that parts within the tolerance allowed by the engineering drawing will be maintained. CNC is now achieving the accuracy, formerly dependent on the machinist's skill with reliable control systems and better machine tool construction.

Reliability

The performance of CNC machine tools and their control systems had to be at least as reliable as their skilled help (machinists, toolmakers, and diemakers, etc.) for industry to accept this new machining concept. Since consumers throughout the world were demanding better and more reliable products, there was a great need for equipment that could machine to closer tolerances and be counted on to repeat this time and time again.

Machine tools were greatly improved when all parts and components were made to closer tolerances. Improvements in machine slides, bearings, ball screws, and machine tables all helped to make sturdier and more accurate machines. New cutting tools and toolholders were developed that matched the accuracy of the machine tool and made it possible to produce accurate parts consistently. The control systems now in use are capable of ensuring that the machine tool will produce accurate parts every time. Factors such as cutter compensation are built into the MCU and make the necessary adjustments to compensate for cutting tool wear, ensuring that accuracy will be maintained.

Repeatability

Repeatability and reliability are very difficult to separate because many of the same variables affect both. Repeatability of a machine tool involves comparing each part produced on that machine to see how it compares to others for size and accuracy. The repeatability of a CNC machine should be at least one-half the smallest tolerance allowed on the part. Machine tools capable of greater accuracy and repeatability will naturally cost more because this accuracy must be built into the machine tool and/or the control system. To maintain the repeatability of a machine tool it is important that a regular maintenance program be established for each machine. Care in loading and clamping parts accurately in workholding devices such as fixtures and vises (whether manually or automatically by robots) is of prime importance, since an incorrectly held or positioned part will result in scrap work.

Productivity

The goal of industry has always been to produce better products at competitive or lower prices to gain a bigger share of the market. Soaring production worldwide has increased the competition for global and domestic markets. To meet this competition, manufacturers must continue to reduce costs and build better-quality products, and they must get greater output per worker, greater output per machine, and greater output for each dollar of capital investment. These factors alone are justification for using CNC and automating factories. It provides us with the opportunity to produce goods of better quality, faster, and at a lower cost.

Let us compare the costs of producing a typical part manually and by CNC. The part shown in Fig. 3-5-4 requires 25 holes of four different sizes, involving the operations of spotting, drilling, countersinking, counterboring, reaming, and drilling. Eleven different cutting tools are required for these operations, and the locational tolerance for the holes is .001 inch (0.025 mm).

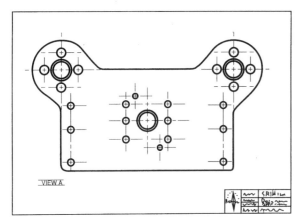

Fig. 3-5-4   A typical part ordered by a customer.
(Coleman Engineering Company)

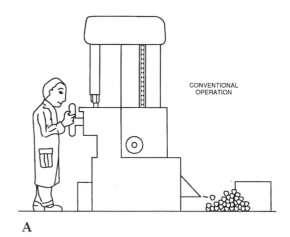

A

Let us assume that a firm received an order for 700 of these parts that had to be supplied, as required by the customer, in lots of 35. Because the delivery date is uncertain, it is quite likely that the firm may produce these only as required, and would need 20 lots or manufacturing runs to complete the order. For a comparison of conventional operation with CNC, see Fig. 3-5-5A and 3-5-5B and Table 3-5-1.

The savings of $8,961 in this example is fairly typical and shows the increased productivity along with savings in production costs. Naturally it is difficult to make a general rule regarding the savings that CNC offers; some jobs will

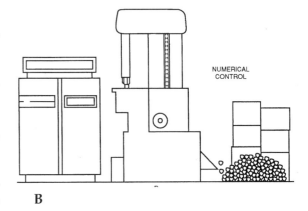

B

Fig. 3-5-5   Producing parts by (A) conventional processes versus (B) CNC which results in increased production rates. *(Coleman Engineering Company)*

Manually	Computer Numerical Control		Savings
2 Fixtures required. Tool design, fabrication, inspection, trial costs = $6,000.00	1 Simple fixture $150.00 Program preparation 2 hrs.@ $60.00/hr. $120.00 = $270.00		$5,730.00
4 – 6 week tooling time	1 day for alignment, fixture, program preparation		
Floor-to-floor time part 30 min. x 700 parts = 350hrs.@$45.00 =15,750.00	Floor-to-floor time/part 10 min. x 700 parts 117 hrs.@ $60.00 = $7,020.00		$8,730.00
Setup time/job lots	Setup time/job lots		
30 minutes x 20 lots =10 hrs. @ $45.00 = $450.00	5 min. x 20 lots = 1.7 hrs. @ $60.00 = $102.00		$552.00
			Total = $ 15,012.00

Table 3-5-1

Fig. 3-5-6 A photoelectric tape reader.
(The Superior Electric Company)

show greater savings, while others will show less.

There are numerous other benefits of CNC besides lower tooling costs and increased production. Some of the more common are:

1. **Reduced space for tool and fixture storage**. CNC tapes and programs take very little space and can be stored in a small cupboard or even a filing cabinet.
2. **Reduced parts inventory.** The CNC program can be reused as often as parts are required. Capital funds are released and storage requirements are reduced.
3. **Lead time** for part design and revisions is greatly reduced.
4. **Uniformity of finished parts.** There is

very little loss for scrap parts because of the accuracy possible through NC.

5. **Inspection time is reduced.** Programs can be pretested and the machine accuracy can be checked periodically.

TAPE READERS

Machine tools that still use NC punched tape need a tape reader to read the coded commands. The reading (decoding) is done by the tape reader mechanism, which is actually a part of the MCU, Fig. 3-5-6. The tape reader decodes the information on the part program from the punched tape and sends it to the MCU, Fig. 3-5-7.

FEEDBACK SYSTEMS

CNC systems can read punched tape or computer programs and direct a machine tool to perform a wide variety of operations. It is important that the machine tool perform these operations correctly and to the exact dimensions called for on the part print. To achieve this accuracy there must be some method of checking the amount a machine table has moved against the amount it was asked to move by the MCU. **Feedback devices** are used to send data back to the MCU so that the machine table slide position can be compared with the required input data dimensions. If any difference exists, the drive

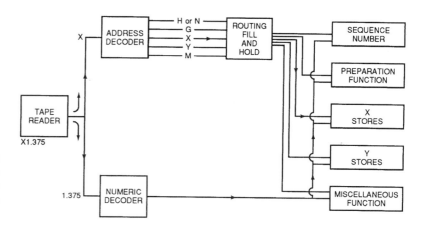

Fig. 3-5-7 A block diagram that shows how the reader decodes the CNC program and sends the information to the proper storage area.
(Kelmar Associates)

Fig. 3-5-8 Analog transducers produce an electrical voltage whose variations in level can be sensed, measured, and converted into accurate distances.

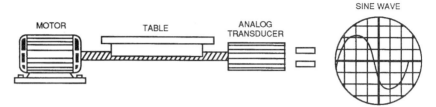

motor will be told (through electrical pulses) to make the necessary corrections.

There are two types of feedback systems, analog and digital. They both use some form of feedback device on the machine slides or leadscrews to indicate the exact position of the machine table. Some of the more common feedback devices are Farrand scales or rotary pickup units.

Analog

Analog transducers, such as potentiometers and synchros, produce an electrical voltage which varies as the input shaft is turned or rotated, Fig. 3-5-8. This voltage is in proportion to the rotation of the input shaft and can be converted into very accurate machine table position information.

Digital

Digital feedback units (rotary resolvers), attached to the leadscrew of a machine tool, trans-

late the rotary motion of the machine screws to individual or discrete electrical pulses, Fig. 3-5-9. This series of pulses can be counted to indicate exactly how much the leadscrew shaft has turned, which indicates the amount the machine table has moved.

SERVO CONTROLS

Servo controls can be any group of electrical, hydraulic, or pneumatic devices that are used to control the position of machine tool slides. The most common servo control systems in use are the open-loop and the closed-loop systems.

Open-Loop System

In the *open-loop system*, Fig. 3-5-10, the tape or program is fed into a *reader* that decodes the program information and stores it briefly until the machine is ready to use it. The tape reader then converts the

(Coleman Engineering Company)

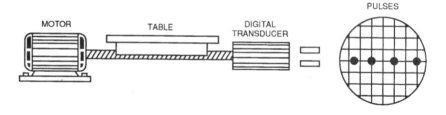

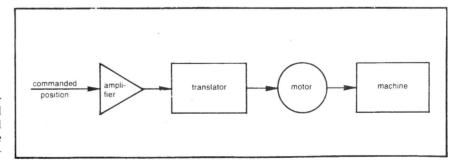

Fig. 3-5-9 Digital transducers produce electrical pulses that can be counted and converted into accurate distances. *(Coleman Engineering Company)*

information into electrical pulses or signals. These signals are sent to the *control unit*, which energizes the *servo control units.* The servo control units direct the *servomotors* to perform certain functions according to the information supplied by the program. The amount each servomotor will move the slide depends upon the number of electrical pulses it receives from the servo control unit. Precision leadscrews, usually having 10 threads per inch (tpi), are used on NC machines. If the servomotor connected to the leadscrew receives 1000 electrical pulses, the machine slide will move 1 inch (25.4 mm). Therefore, one pulse will cause the machine slide to move .001 inch (0.0254 mm.). The open-loop system is fairly simple; however, there is no means of checking whether the servomotor has performed its function correctly so it is not generally used where an accuracy greater than .001 inch (0.025 mm) is required.

The open-loop system may be compared to a gun crew that has made the calculations necessary to hit a distant target but does not have an observer to confirm the accuracy of the shot.

The *closed-loop system,* Fig. 3-5-11, can be compared to the same gun crew that now has an observer to confirm the accuracy of the shot. The observer relays the information regarding the accuracy of the shot to the gun crew, which then makes the necessary adjustments to hit the target. The closed-loop system is similar to the open-loop system with the exception that a *feedback*

unit, Fig. 3-5-11, is introduced into the electrical circuit. This feedback unit, often using a rotary resolver or *transducer,* compares the amount the machine table has been moved by the servomotor with the signal sent by the control unit. The control unit instructs the servomotor to make whatever adjustments are necessary until both the signal from the control unit and the one from the servo unit are equal. In the closed-loop system, 10,000 electrical pulses are required to move the machine slide 1 inch (25 mm). Therefore, on this type of system, one pulse will cause a .0001-inch (0.0025-mm) movement of the machine slide. Closed-loop NC systems are very accurate because the command signal is recorded, and there is an automatic compensation for error. If the machine slide is forced out of position due to cutting forces, the feedback unit indicates this movement and the machine control unit (MCU) automatically makes the necessary adjustments to bring the machine slide back to position.

MCU

The MCU, Fig. 3-5-12, is the intermediary in the total NC operation. Its main function is to store the part program in a language that the machine tool can understand so that it can perform the func-

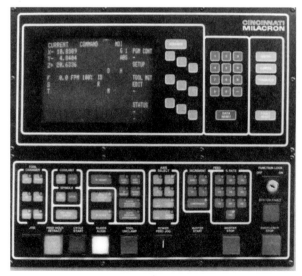

Fig. 3-5-11 A closed–loop system checks the accuracy of a move with the input dimensions.
(Coleman Engineering Company)

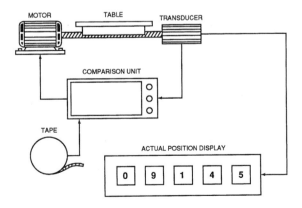

Fig. 3-5-10 Open–loop CNC systems do not have any means of checking the accuracy of a move. *(Kelmar Associates)*

tions required to produce a finished part. These functions could include turning relays or solenoids ON or OFF and controlling the machine tool movements through electrical or hydraulic servomechanisms.

Data Decoding and Control

One of the first operations that the MCU must perform is to take the binary coded data (BCD) from the program and change it into binary digits. This information is then sent to a holding area, generally called the buffer storage, of the MCU, Fig. 3-5-13. The purpose of the buffer area is to allow the information or data to be transferred faster to other areas of the MCU. If there were no buffer storage, the MCU would have to wait until the reader decoded and sent the next set of instructions. This waiting would cause slight pauses in the transfer of information, which in turn would result in a pause in the machine tool motion and cause tool marks in the workpiece. MCUs which do not have buffer storage must have high-speed readers to avoid the pauses in transferring information and the machining operation.

The data decoding and control area of the MCU processes information which controls all machine tool motions as directed by the program. This area also allows the operator to stop or make changes to the program manually, through the control panel.

MCU Development

Since the early 1950s, MCUs have developed from the bulky vacuum tube units to today's computer control units, which incorporate the latest microprocessor technology. Until the early 1970s, all MCU functions—such as tape format recognition, absolute and incremental positioning, interpolation, and code recognition—were determined by the electronic elements of the MCU. This type of MCU was called *hard-wired* because the functions were built into the computer elements of the MCU and could not be changed.

The development of *soft-wired* controls in the mid-1970s resulted in more flexible and less costly MCUs. Simple types of computer elements, and even minicomputers, became part of the MCU. The functions that were locked in by the manufacturer with the early hard-wired systems are now included in the computer software within the MCU. This *computer logic* has more capabilities, is less expensive, and at the same time can be programmed for a variety of functions whenever required.

CPU

The CPU of any computer contains three sections: control, arithmetic-logic, and memory. The CPU control section is the workhorse of the computer, Fig. 3-5-14. Some of the main functions of each part of the CPU are as follows:

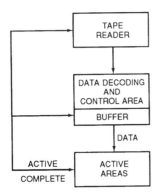

When active areas complete current block, new data are transferred from buffer, and the tape reader starts to refill the buffer with new tape data

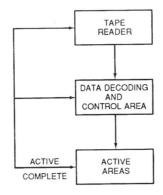

When active areas complete current block, the tape reader starts to read into the active areas the new data

Fig. 3-5-12 The MCU controls the flow of data in a CNC control system. *(Cincinnati Machine, aUNOVA Co.)*

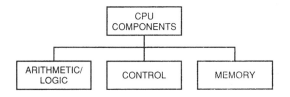

Fig. 3-5-13 The buffer storage quickly transfers infor-

1. **Control section**
   — Coordinates and manages the whole computer system.
   — Obtains and decodes data from the program held in memory.
   — Sends signals to other units of the CNC system to perform certain operations.
2. **Arithmetic-Logic**
   — Makes all calculations, such as adding, subtracting, multiplying, and counting as required by the program.
   — Provides answers to logic problems (comparisons, decisions, etc.).
3. **Memory**
   — Provides short-term or temporary storage of data being processed.
   — Speeds the transfer of information from the main memory of the computer.
   — Has a memory register that provides a specific location to store a word and/or recall a word.

COMPUTER NUMERICAL CONTROL

In 1970, a new term—computerized numerical control (CNC)—was introduced to the NC vocabulary. The development of a CNC system was made economically possible by electronic breakthroughs that resulted in lower costs for minicomputers.

The physical components of CNC or *soft-wired* units are the same regardless of the type of machine tool being controlled. On soft-wired units, it is not the MCU but the *executive program,* or *load tape,* which is loaded into the CNC computer's memory by the control manufacturer that makes the control unit "think" like a turning center or a machining center. If there is a need to change the functions of a soft-wired or CNC unit, the executive program can be changed to suit. Usually it is the manufacturer and not the user, who makes the change.

There have been great developments in MCUs with the introduction of large-scale ICs and microprocessors. Many features not available a few years ago give much greater flexibility and productivity to NC systems. As developments in control units continue, new features will become available that will make the machines they control more productive.

The CNC machine control unit of today has several features that were not found on the pre-1970 hard-wired control units, Fig. 3-5-15.

1. **Cathode-Ray Tube (CRT)**
   The CRT is like a TV screen and serves a number of purposes:
   a) It shows the exact position of the machine table and/or the cutting tool at every position while a part is being machined.
   b) The program for a part can be displayed on the screen for editing or revision purposes.
   c) The screen assists in work setups, and some models receive messages from sensors that indicate machine or control problems.
2. **Absolute and Incremental Programming**
   By using the proper address code G90 (absolute) or G91 (incremental), CNC units will automatically program in that particular mode. CNC units are capable of handling mixed data (absolute and incremental) in a given program.
3. **Inch or Metric**
   CNC units are capable of working in inch or metric dimensions. Either a switch on the control unit or a specific code on the part program (G20 for inch and G21 for metric) will determine the measurement system used when machining a part.
4. **EIA or ASCII Code**
   CNC units will read either the Electronic Industries Association (EIA) or the Amer-

ican Standard Code for Information Interchange (ASCII) standard code. The control unit identifies each one by the parity check: odd for the EIA code and even for the ASCII code.

5. **Manual Data Input**

   CNC units provide some method of making changes to the part program, if necessary. This may be necessary because changes were made to the part specifications, to correct an error, or to change the machining sequences of the part.

6. **Program Editing**

   Very few part programs are free from error from the start, and the flaws generally show up on the shop floor. Program editing is a feature that allows the part program to be corrected or changed at the control unit.

7. **Interpolation**

   Early models of control units were capable of only linear, circular, or parabolic interpolation, but the newer models include helical and cubic interpolation.

8. **Point-to-Point and Continuous-Path Positioning**

   All MCUs are capable of point-to-point or continuous-path positioning or any combination of each.

9. **Cutter Diameter and Length Compensation**

   With today's MCUs, it is possible to manually enter the diameter and/or the length of a cutter which may vary from the specification of the part program. The control unit automatically calculates the adjustments necessary for the differences in size and moves the necessary slides to adjust for them.

10. **Program Storage**

    The newer CNC units generally provide large-capacity computing and data storage (memory). This arrangement allows information about a part program to be entered (manually or from the program) and stored for future use. Whenever the part program is required it can be recalled from memory rather than reread from the program. This facility not only protects the quality of the program, but the information is recalled much faster.

10. **Canned or Fixed Cycles**

    Storage capacity is generally provided in the MCU for any cycle (machining, positioning, etc.) that is used or repeated in a program. When the program is being written and a previous cycle must be repeated, it is only necessary to insert a code in the program where that cycle is re-

mation to other areas of the MCU to ensure continuous machine tool motion.

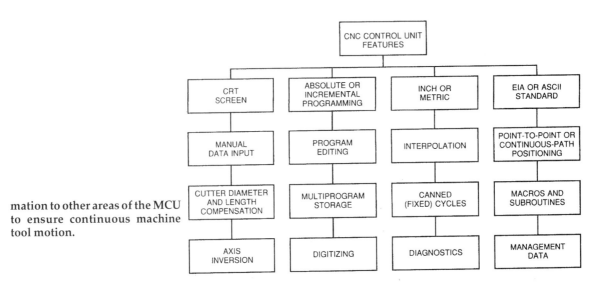

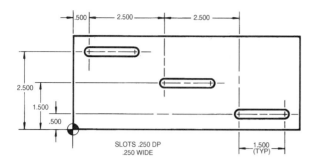

Fig. 3-5-14 The control functions of the computer CPU.
(Kelmar Associates)

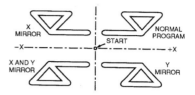

Fig. 3-5-15 Today's CNC units contain many more control features than were found on earlier control units. *(Kelmar Associates)*

quired. The control unit will recognize that code and recall from memory all the information required to repeat that cycle (operation).

12. **Subroutines and Macros**

A parametric subroutine, sometimes called a "program within a program," is used to store frequently used data sequences (one block or a number of blocks of information) which can be recalled from memory as required by a code in the main part program, Fig. 3-5-16. An example of a subroutine could be a milling cycle in which a series of slots .250 inch wide (6.35 mm) must be machined in a number of locations on a workpiece.

A *macro is* a group of instructions or data that are permanently stored and can be recalled as a group to *solve recurring problems* such as bolt-hole circle locations, drilling and tapping cycles, and other frequently used routines. An example of a macro would be the XY locations of various holes on any bolt-circle diameter. When the diameter of the bolt circle and the number of holes on the circle are provided, the MCU makes all the calculations for hole locations and causes the machine tool slides to move into the proper position for each hole.

13. **Axis Inversion (Mirror Image)**

Axis inversion, Fig. 3-5-17, is the ability of the MCU to reverse plus and minus (+ and –)

values along an axis or program zero to produce an accurate left-hand part from a right-hand program. This ability (called *symmetrical machining*) applies to all four quandrants and greatly reduces the time that would be required to program each part.

14. **Digitizing**

The digitizing feature allows a part program to be made directly from an existing part. The original part is traced while the CNC unit records all machine motions and automatically produces them on the part program.

15. **Diagnostics**

Diagnostic capabilities (built-in) can monitor all conditions and functions of a CNC machine or the control unit. If an error or malfunction occurs or if a changing condition nears a critical point, a signal or message is shown on the CRT. On some control units, as a critical point gets close, there may be a warning to the operator, shown on the screen, or the machine may automatically shut down.

Diagnostic software routines can be used to check hardware modules, circuit boards, and every area of the control unit to check their accuracy. The information is then displayed on the CRT.

16. **Management Data**

The modern MCU controls almost all machine tool functions through the built-in computer. Since this information or data is already in the MCU controller or computer, it can be sent to the host or mainframe computer to provide valuable data to management and the operator. Spindle-on time, part-run time, number of parts machined, etc., can be recorded and sent to the host computer or displayed on the CRT screen.

ADVANTAGES OF NC

Recent studies show that of the amount of time an average part spent in a shop, only a fraction of that time was actually spent in the machining process. Figure 3-5-18 shows the distribution of time in a shop where parts are machined in small batches. Let us assume that a part spent 50 hours from the time it arrived at a plant as a rough casting or bar stock to the time it was a finished product. During this time, it would be on the machine for only 2 1/2 hours and be cut for only 3/4 hour. The rest of the time would be spent on waiting, moving, setting up, loading, unloading, inspecting the part, setting speeds and feeds, and changing cutting tools.

CNC reduces the amount of non-chip-producing time by selecting speeds and feeds, making rapid moves between surfaces to be cut, using automatic fixtures, automatic tool changing, controlling the coolant, in-process gauging, and loading and unloading the part. These factors made it unnecessary to train machine operators, these processes have resulted in considerable savings throughout the entire manufacturing process and tremendous growth in the use of CNC. Some of the major advantages of CNC are as follows:

1. There is automatic or semiautomatic operation of machine tools. The degree of automation can be selected as required.
2. Flexible manufacturing of parts is much easier. Only the program needs to be changed to produce another part.
3. Storage space is reduced. Simple workholding fixtures are generally used, reducing the number of jigs or fixtures that must be stored.
4. Small part lots can be run economically. Often a single part can be produced more quickly and better by CNC.
5. Nonproductive time is reduced. More time is spent on machining the part, and less time is spent on moving and waiting.
6. Tooling costs are reduced. Complex jigs and fixtures are generally not required.
7. Inspection and assembly costs are lower. The quality of the product is improved, reducing the need for inspection and ensuring that parts fit as required.
8. Machine utilization time is increased. A machine tool is idle for less time because workpiece and tool changes are rapid and automatic.
9. Complex forms can easily be machined. The new control unit features and programming capabilities make the machining of contours and complex forms very easy.
10. Parts inventory is reduced. Parts can be made as required from the information on the program.

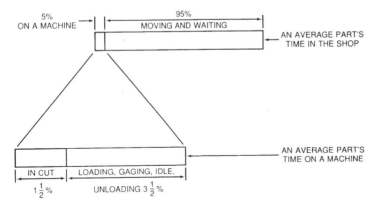

Fig. 3-5-16 Subroutines consist of a block of information that is stored in memory and recalled when a similar operation must be repeated.
(*The Superior Electric Company*)

Since the first industrial revolution, about 200 years ago, CNC has had a significant effect on the industrial world. Developments in the computer and CNC have extended a person's mind and muscle. The CNC unit takes symbolic input and changes it to useful output, expanding a person's concepts into creative and productive results. CNC technology has made such rapid advances that it is being used in almost every area of manufacturing, such as machining, welding, pressworking, and assembly. If industry's planning and logic are good, the second industrial revolution will have as much or more effect on society as the first industrial revolution. As we progress through the various stages of CNC, it is the entire manufacturing process that must be kept in mind. Computer-assisted manufacturing (CAM), computer-integrated machining (CIM), and virtual reality (VR) are certainly where the future of manufacturing lies, and because of the developments in the past, the automated factory has become a reality.

SUMMARY

- The CNC system is a linking element between the operator (operator interface) and the machine tool (machine tool interface).
- The development of transistors, integrated circuits, and the computer chip made the development of CNC possible.
- CNC is widely accepted throughout the world because of its accuracy, reliability, repeatability, and productivity.
- The closed-loop feedback system is the most accurate because it compares the amount a machine slide moves to the amount it was instructed to move and makes corrections whenever necessary.
- The function of a MCU that contains the control system, arithmetic logic, and memory section, is to decode programming language into language a machine tool can understand.
- CNC has made it possible to manufacture parts faster and more accurately, increased machine utilization time, reduced parts inventory and inspection costs, and made flexible manufacturing systems possible.

KNOWLEDGE REVIEW

1. Name three key elements of a CNC control system.

CNC Performance

2. List four important developments that had an important effect on the great advances made by CNC.
3. To what accuracy are CNC machine tools capable of working? Explain why.
4. What improvements have been made to machine tools to make them more reliable?
5. How were control systems improved to ensure that accurate products were produced?
6. Define the term *repeatability*.
7. Name three factors that alone are justification for using CNC.

8. Compare the manual machining of a part with CNC machining with regard to fixtures, floor-to-floor time, setup time, and dollar savings.

Tape Readers

9. What is the purpose of a tape reader?

Feedback Systems

10. What is the purpose of feedback devices?
11. Name two types of feedback systems and state how each operates.

Servo Controls

12. Briefly explain how an open-loop system operates.
13. Compare the closed-loop system with the

open-loop system and state its advantages.

14. How many electrical pulses are required to move a machine slide 1 inch (25 mm) on a closed-loop system?

MCU

15. What is the purpose of the MCU?
16. What is the purpose of the data decoding and control section of the MCU?
17. Explain the difference between hard-wired and soft-wired controls.
18. What purposes does the CPU serve?
19. List two important functions of each of the CPU sections.

CNC

20. Explain the purpose of the executive program.
21. What is the purpose of the CRT?

22. How can the MCU recognize absolute or incremental programming?
23. What is the purpose of cutter diameter and length compensation?
24. Briefly explain:
    (a) Canned or fixed cycles
    (b) Subroutines and macros
    (c) Digitizing
25. What is the purpose of the axis inversion feature?

Advantages of NC

26. If an average part spends about 50 hours in a shop, how much time is likely to be spent in actually machining the part? How was the rest of the time most likely spent?
27. Why has CNC been so widely accepted by industry?
28. List five of the most important advantages of CNC.

SECTION 4

Programming Data

Unit 6 CNC Functions

PROGRAMMING DATA

CNC machine tools need numerical data in order to control the motion (relationship) between the cutting tool and the workpiece in accordance with the dimensions on a part print. Other numerical data can set feeds and speed rates, define tool identification numbers, and miscellaneous functions related to machining the part. The combination of all the numerical data, in a sequence understood by the machine CNC controller, is called the part program.

Computer Numerical Controls use standard microelectronics modules developed especially for computer hardware. CNC controllers, through one or more microprocessors, can perform the mathematical calculations necessary for complicated moves. Modern CNCs can store a large number of programs and provide displays of tool paths and simulation of the machining operation.

The programmer's function is to take information from an engineering drawing and convert it into data that the control unit of the machine tool will understand. To do this conversion successfully, a programmer must have a good knowledge of machining operations and sequences, workholding methods, metal properties, cutting tools, and cutting speeds and feeds. The programmer must also understand the various codes and functions used in computer numerical control (CNC) so that accurate parts can be produced in the minimum amount of time. It is very important that there be a close liaison between the product engineer, the drafter, and the machine tool operator to ensure that the correct product is produced to the proper size.

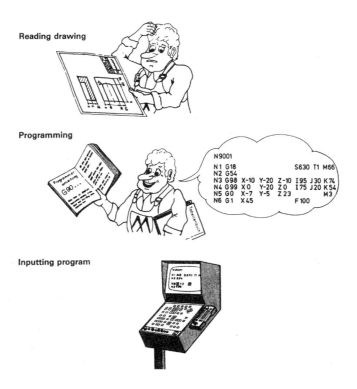

Unit 6
CNC Functions

The digital encoding of data is necessary to reduce storage space, and provide high-speed data transmission, and fast retrieval of data from the input media. It is a fact that electronic devices cannot think for themselves and that they do what they have been programmed to do by humans. Therefore it is necessary to accurately prepare the CNC program and check it for accuracy before it is used on a machine.

OBJECTIVES

After completing this unit, you should be able to:

1. Understand the functions of CNC preparatory and miscellaneous codes.
2. Describe the order in which information is generally found in a block of information.
3. Know the purpose and function of the common fixed or canned cycles.
4. Identify the correct speeds and feeds to be used for cutting various metals.

KEY TERMS

address character	machine zero
axis movement	motion axes
canned cycles	sequence number
gage height	work plane

CNC FUNCTIONS

The Electronic Industries Association (EIA) and the American Standard Code Information Interchange (ASCII) have standardized CNC functions for programming. Most CNC machine and control manufacturers use these coding systems. Although there are slight variations between each system, the function codes are basically the same. Various types of function codes are used in CNC work to indicate the type of operation that the machine tool is to perform. The most common are the *preparatory* or G functions, the *miscellaneous* or M functions, and the *address character* or A to Z functions, which must appear on the manuscript, and the CNC program.

CNC programming is done in a variable-block format with word (letter) addressing. Each instruction word consists of an address character (X, Y, Z, F, or M) followed by the numerical data. Fig. 4-6-1. The letter address (code) tells the machine control unit (MCU) what machine function to

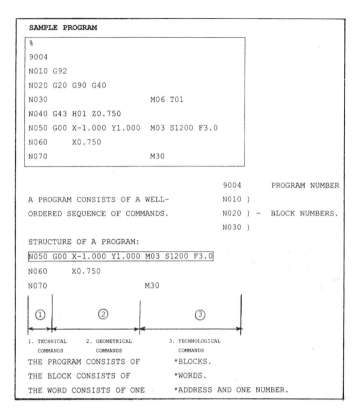

```
SAMPLE PROGRAM

%
9004
N010 G92
N020 G20 G90 G40
N030                    M06 T01
N040 G43 H01 Z0.750
N050 G00 X-1.000 Y1.000  M03 S1200 F3.0
N060      X0.750
N070                    M30


                              9004      PROGRAM NUMBER

A PROGRAM CONSISTS OF A WELL-     N010 )
ORDERED SEQUENCE OF COMMANDS.     N020 ) - BLOCK NUMBERS.
                                  N030 )

STRUCTURE OF A PROGRAM:

N050 G00 X-1.000 Y1.000 M03 S1200 F3.0
N060      X0.750
N070                    M30

  ①          ②              ③

1. TECHNICAL   2. GEOMETRICAL   3. TECHNOLOGICAL
   COMMANDS       COMMANDS         COMMANDS
THE PROGRAM CONSISTS OF   *BLOCKS.
THE BLOCK CONSISTS OF     *WORDS.
THE WORD CONSISTS OF ONE  *ADDRESS AND ONE NUMBER.
```

Fig. 4-6-1 A CNC program consists of a sequence of commands that contain blocks, words, and letter addresses. *(Kelmar Associates)*

perform, and the number code usually gives the distance, feed rate, speed, etc. Let us examine a sample program of CNC information in order to understand programming sequence.

BLOCK FORMAT

The structure of CNC programs consists of a number of well-ordered sequences of commands, Fig. 4-6-1. Each program command can consist of:

• Technical commands such as program numbers (N010, N020, etc.) and block numbers (labeled N, H, or O).
• Geometrical commands, such as G01, for distances in the X, Y, and Z axes or G02 for circular interpolation.
• Technological commands or miscellaneous functions such as speed (S), feed (F), spindle ON or OFF, etc.

A brief description of some of the more common program commands used in CNC follows:

1. Block number—consists of the letter (N) followed by up to 5 digits; N _ _ _ _ _.
2. G-address (preparatory functions)- consists of the letter followed by up to 2 digits, G _ _.
3. X, Y, Z addresses (dimensioning codes)— have the letter, plus or minus (±) signs, and 8 digits.
   Vertical machine: X ±, Y ±, Z ± 8 digits.
   Horizontal machine: X ±, Y ±, Z ± 8 digits.
   Turning machine: X ±, Z ±, 8 digits.
4. F (feed) address—The feed rate of the cutting tool, stated as a letter and up to three digits; F _ _ _.
5. S (speed) address—generally refers to spindle speed, consists of a letter and up four or five digits, S _ _ _ _.
6. M-address (miscellaneous functions)— consists of a letter followed by up to 2 digits; M _ _.
7. T-address (tool number)-consists of a letter followed by up to three digits; T _ _ _.

8. J,K-addresses (circle parameters)—consist of a letter followed and up to 8 digits, J _ _, K _ _.

Most CNC manufacturers follow the **word-addresses format** listed below for positioning and contouring machines. The earlier tab sequential or fixed formats virtually disappeared in the 1970s and are now considered obsolete.

A—Rotation around X axis
B—Rotation around Y axis
C—Rotation around Z axis
F—Feed rate
G—Preparatory functions
I—Circular interpolation X axis offset
J—Circular interpolation Y axis offset
K—Circular interpolation Z axis offset
M—Miscellaneous functions
N—Sequence or block number
R—Arc radius
S—Spindle speed
T—Tool number
X—X axis data
Y—Y axis data
Z—Z axis data
%—Program start (rewind)

See the example of the block format and size shown in blocks N040 and N050 in Fig. 4-6-1 for information on how various addresses are used in a CNC program. There are differences in programming for CNC machining and turning centers, and it is important to be familiar with the programming for each machine.

SEQUENCE NUMBER

The first word in a block of information is usually the *sequence number,* and it is used to identify each particular block of information. The sequence number is usually a three-digit numerical word (though it can also be four digits) preceded by the letter code N.

- The letter N is generally used to identify all other sequence blocks. Sequence numbers range from N1 to N99999.
- Sequence numbers are especially valuable

when it is necessary to edit and make corrections or revisions to a program.
- The program can be made to search for a particular sequence number, stop when it is found, and display it on the cathode-ray tube (CRT) screen for editing purposes.
- If sequence numbers are assigned with flexibility in mind during programming, they can be very useful if it becomes necessary to add, revise, or correct a program.
- It is not considered good practice to assign numbers in numerical order (for example, N001, N002, N003), because this approach provides no flexibility if revisions or additions must be made to a program. If a correction or revision must be made to the program, all following sequence numbers must be renumbered.
- The most common method used to assign sequence numbers is in a progression of 10 (for example, N010, N020, N030). This arrangement allows room between each sequence number to insert as many as nine pieces of new information before any sequence numbers would have to be renumbered. The following illustrates the numerical order approach starting with number 1.

N001 New sequence inserted must become N002.
N002 All following sequences must be renumbered.

The following illustrates the progression of 10 system.

N010 New sequences N011 to N019 can be added.
N020 No renumbering required.
N030

MACHINE MOTION AXES

All machine tools have sliding linear and rotating motion. The single-spindle CNC machine, Fig. 4-6-2, has three linear motions (two horizontal and one vertical) and one rotary motion (the spindle). CNC makes it very important to identify these motions accurately so that they can be

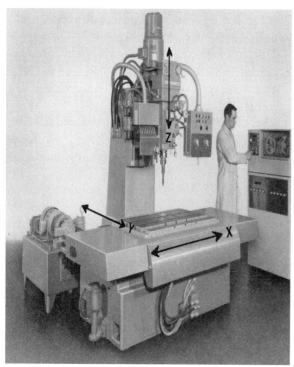

Fig. 4-6-2 The X, Y, and Z axes of a CNC machine give three–dimensional coordinates, with the Z axis in a vertical relationship to the XY axes.
(Cincinnati Machine, a UNOVA Co.)

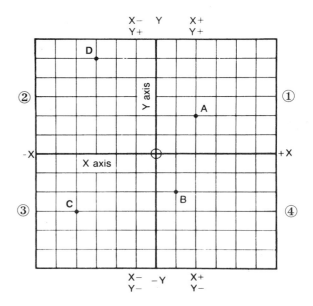

Fig. 4-6-3 Locating points within the XY coordinates. *(Allen–Bradley)*

called the X axis, and the vertical line is called the Y axis. To avoid confusion and errors when programming, the distance to the right or left of the Y axis is generally given first.

- Any X distance to the right of the Y axis is referred to as a positive (+) dimension; those to the left are negative (–) dimensions.
- Any Y distance above the X axis is referred to as a positive (+) dimension; those below are negative (–) dimensions.

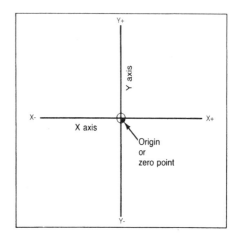

Fig. 4-6-4 Intersecting horizontal and vertical lines form a right angle. *(Allen–Bradley)*

programmed and dimensioned as required. The rectangular coordinates (X, Y, and Z) make it possible to locate any point on a single plane or in a three-dimensional space.

The location of any point on a flat surface can be positioned accurately by making reference to where it is in relation to the X and Y axes, Fig. 4-6-3. The rectangular coordinates also allow any point in space to be located in relation to three perpendicular planes (XYZ) axes.

X, Y, and Z Words

X, Y, and Z words refer to coordinate movements of the machine tool slides for positioning or machining purposes. When points are located on a workpiece, two intersecting lines (one horizontal and the other vertical) at right angles to each other are used. These lines are called "axes," and where they intersect is called the "origin," or "zero point", Fig. 4-6-4. The horizontal line is

• The third plane, or Z axis, is perpendicular to the plane established by the X and Y axes and on a vertical mill or drill press refers to the movement of the cutting tool.
— Toward or into the workpiece is a negative (–) motion, and movement away from the workpiece is a positive (+) motion.

Coordinate information must be programmed in the following sequence in order for the MCU to understand the command correctly.

1. *Axis Movement*
   X, Y, or Z axes must be specified. When motion is required on more than one axis, program X first, Y second, and Z last.
2. *Direction Movement*
   The information must indicate whether the movement is positive or negative (–) from the origin point.
   • The positive movement sign (+) does not have to appear in a program; if no sign is entered it is automatically assumed that it is positive.
   • Negative dimensions must have the minus (–) sign. For example, the dimension X6.875 would mean a 6 inch and 875 thousandths of-an-inch movement to the right of the Y axis.
3. *Dimension Movement*
   Normally a seven-digit number with the decimal point fixed in the program format to allow four places to the right of the decimal.
   • All modern programming uses decimal point for dimensional input.
   • The control system that supports programming the decimal point can also accept dimensional values without the decimal point to allow compatibility with older systems.
   NOTE: It is important to be aware of the principles of programming format using leading and following zeros or there could be serious problems with the program.
4. Depth Selection (Z Axis)
   Machining operations such as producing holes, milling slots or steps, or cutting into

the surface of a workpiece generally involve the Z axis. The Z axis on any machine tool is usually a line drawn through the center of the machine spindle. Figure 4-6-2 shows the Z axis in relation to the X and Y axes of the machine table and the workpiece.

• The Z axis of motion is always parallel to the spindle of the machine and perpendicular to the workholding surface.
— On milling, boring, drilling and tapping machines, the spindle is the tool-rotating device.
— On lathes, grinders, and other machines where the work revolves, the spindle is the work-rotating device.
• A positive Z(+) movement moves the cutting tool *away* from the workpiece, while a negative Z (–) movement moves the cutting tool *into* a workpiece.
• The Z motion can be controlled by the operator (manual data input), or by the computer program.

The word-address letter Z generally consists of a seven-digit number with the decimal point fixed at four places to the left. The Z movement generally involves a rapid move to a programmed gage height or a position above the workpiece and then a slower feed rate during the machining operation. The following formula can be used if the entire Z axis movement must be programmed.

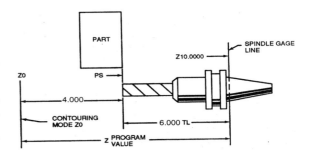

Fig. 4-6-5 The complete axis movement starts at spindle gage line and includes the clearance, if necessary. (*Cincinnati Machine, aUNOVA*)

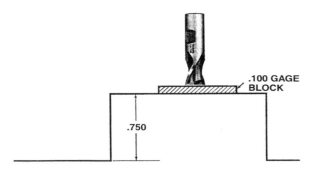

.100 GAGE
BLOCK

.750

Fig. 4-6-6 Establishing the gage height or R 00 work surface location. *(Kelmar Associates)*

Z =   PS + CL + TL
Z =   the distance from Z0 (zero) to the spindle gage line
PS =   the distance from Z0 to the part surface
CL =   the clearance
TL =   the tool set length (from the spindle gage line to the cutting edge)

Work Plane

In a fixed cycle, the word address letter **R** refers to either the work surface or the rapid-feed distance (sometimes called the work *plane)* programmed. The **R** work surface is set either at the highest surface on the part to be machined or at a specific height or distance above this surface. This surface setting is referred to as **R** 00, or the reference dimension, and all programmed depths for cutting tools and surfaces to be machined are taken from the **R** surface.

The **R** work plane (**R** 00) is generally established at .100 inch above the highest surface of the workpiece, Fig. 4-6-6. This surface is also known as gage height. When cutting tools are set up, the operator generally places a .100 inch thick gage on top of the highest surface of the workpiece, and all tools (regardless of length) are set to this height. Once the gage height has been set, it is not generally necessary to add this distance (.100 inch) when changing work surfaces, since most MCUs automatically add the .100 inch to all future depth dimensions.

PREPARATORY FUNCTIONS

The *preparatory function* or *cycle code* refers to a mode of operation of the machine tool or CNC system. It generally refers to some action on the X, Y, and/or Z axes.

- In CNC programming, the word address letter G refers to a preparatory function and is followed by a two-digit number; e.g., a G00 function would be point-to-point positioning at a rapid rate of about 150 to 400 inches/min (38 to 100 m/min).
- Common preparatory functions include operations such as point-to-point positioning, linear interpolation, parabolic interpolation, absolute or incremental programming, inch or metric programming, and fixed (canned) cycles. Each of these is designated by a G code, which the central processing unit (CPU) and the machine tool can recognize and act upon.

The most commonly-used preparatory functions in Fig. 4-6-7 have been supplied by the EIA and are listed in their standard EIA-274-D.

FIXED OR CANNED CYCLES

Fixed or *canned cycles,* identified by the preparatory function codes G81 to G89, are a *preset combination of operations* which cause the machine axis movement and/or cause the machine spindle to complete such operations as drilling, boring, and tapping. Each operation may involve as many as seven machine movements, and a G81 to G89 code is all that is required on the computer program to make the machine perform a particular operation. Control units with this feature can save up to 50 percent in programming time and one third of the data processing time, and reduce the length of the program.

Most manufacturers of CNC machines and control systems use the standard EIA numbering systems for fixed or canned cycles. A few of the most common cycles are discussed in this unit.

Group	G code	Function
01	G00	Rapid positioning
01	G01	Linear interpolation
01	G02	Circular interpolation clockwise (CW)
01	G03	Circular interpolation counterclockwise (CCW)
00	G04	Dwell
00	G10	Offset value setting
02	G17	*XY* plane selection
02	G18	*ZX* plane selection
02	G19	*YZ* plane selection
06	G20	Inch input (in.)
06	G21	Metric input (mm)
00	G27	Reference point return check
00	G28	Return to reference point
00	G29	Return from reference point
07	G40	Cutter compensation cancel
07	G41	Cutter compensation left
07	G42	Cutter compensation right
08	G43	Tool length compensation in positive (+) direction
08	G44	Tool length compensation in minus (−) direction
08	G49	Tool length compensation cancel
09	G80	Canned cycle cancel
09	G81	Drill cycle, spot boring
09	G82	Drilling cycle, counterboring
09	G83	Peck drilling cycle
09	G84	Tapping cycle
09	G85	Boring cycle #1
09	G86	Boring cycle #2
09	G87	Boring cycle #3
09	G88	Boring cycle #4
09	G89	Boring cycle #5
03	G90	Absolute programming
03	G91	Incremental programming
00	G92	Setting of program zero point
05	G94	Feed per minute

Fig. 4-6-7 Commonly used milling preparatory function codes according to Standard EIA–274–D.

Drill Cycle (G81) (Fig. 4-6-8)

In sequence number (N040), a drill cycle (G81) is used to drill a .500 dia. Hole 1.00 in. deep (Z1.10) at 6 in./min at position #1 (X5.000 Y3.500)

1. The spindle "rapids" to gage height at level #1; then it feeds at 6 in./min to depth at level #2 (Z1.10).
2. The spindle rapids back to gage height (level #1).
3. In sequence N050, the table moves (X7.500) along the X axis to position #2 and repeats the drill cycle.

To calculate the programmed Z depth (level #2) for a 118° included angle drill point, use the following formula:

$$Z = \text{Full body depth} + \text{drill point length}$$
$$= 1.000 + (.300 \times \text{drill diameter})$$
$$= 1.000 + (.300 \times .500)$$
$$= 1.000 + .150$$
$$= 1.150$$

Drill/Dwell Cycle (G82) (Fig. 4-6-9)

In sequence N060, code G82 calls for a dwell cycle. This cycle is the same as a drill cycle with a dwell time added at the programmed depth.

1. The spindle feeds from the gage height (level #1) to the depth at level #2.
2. The spindle stops at the depth for a period of time that is preset on the dwell timer, and then it finishes the cycle.
3. In sequence N070, the table moves along the X axis (X5.000) to position #2.
4. The cycle is repeated at position #2.

Note: The dwell cycle is generally used for operations such as counterboring and spot facing where a smooth surface is required.

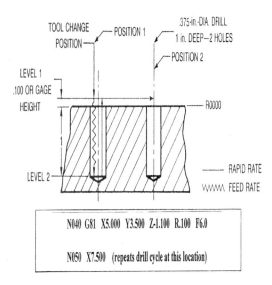

Fig. 4-6-8 A drill cycle to drill two .500 in. diameter holes 1.00 in. deep. *(Cincinnati Machine, a UNOVA Co.)*

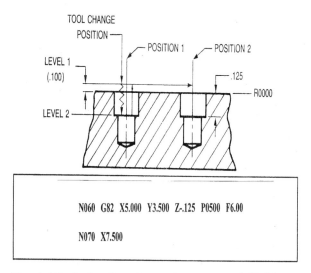

Fig. 4-6-9 A dwell cycle produces a good finish on holes requiring spot facing or counterboring. *(Cincinnati Machine, a UNOVA Co.)*

Tap Cycle (G84) (Fig. 4-6-10)

In sequence N080, code G84 calls for a tap cycle to tap a hole .875 in. deep (Z.875) at 25.5 in./min feed rate.

1. The spindle feeds from gage height at level #1 to the depth at level #2.

2. At .875 in. depth, the spindle reverses and feeds back up to gage height (level #1).

3. The spindle then reverses direction again.

4. In sequence N090, the table moves to 7.500 along the X axis to position #2, and the next hole is tapped.

Bore Cycle (G85) (Fig. 4-6-11)

In sequence N100, code G85 calls for a bore cycle to bore a hole 1.250 in. deep at a feed rate of 2 in./min.

1. The table positions in the X and Y axes, and then the spindle rapids to level #1.

2. The boring tool is then fed to depth at level #2.

3. The direction of feed reverses, and the boring tool returns to level #2.

4. In sequence N110, the table moves along the X and Y axes to position #2, and the cycle is repeated.

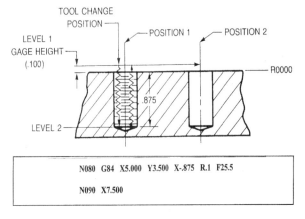

Fig. 4-6-10 A tap cycle used to thread two holes 1.00 in. deep. *(Cincinnati Machine, a UNOVA Co.)*

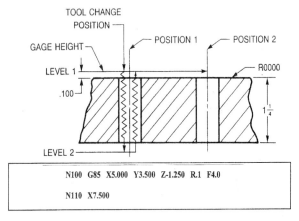

Fig. 4-6-11 The bore cycle is used to bring a hole to accurate size and location.
(Cincinnati Machine, a UNOVA Co.)

ZERO SUPPRESSION

Some machine actuation registers accept word address information in a right-to-left sequence with the decimal point fixed at four places to the left. There are eight digit positions available, so dimensions as large as 999.9999 inches can be programmed. If a dimension of X0068750 was programmed, it would fill the register as shown in Fig. 4-6-12.

Some controls allow all leading zeros—those before the whole number(s) to the left of the decimal point—to be omitted. The words enter the register from right to left, and therefore the two leading zeros before the number 6 have no value and can be suppressed or omitted. This convention is called leading zero suppression. The dimension in Fig. 4-6-12 would be entered as X6.875.

Other controls allow all trailing zeros—those after the number(s) to the right of the decimal point—to be omitted. The words enter the register from left to right. Therefore the one trailing zero after the number 5 has no value and can be suppressed or omitted. The elimination of insignificant zeros is called trailing zero suppression. The dimension in Fig. 4-6-12 would be entered as X6.875.

FEEDS AND SPEEDS

Feed

Feed is the amount that a cutting tool advances into the work, which generally controls the amount and rate of metal removal from a workpiece. The feed rate is generally measured in inches per revolution (in/rev) or in inches per minute (in./min).

- On most CNC machine tools, the rapid feed

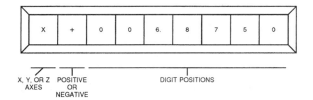

Fig. 4-6-12 The location of the decimal point and digits in a standard register. *(Kelmar Associates)*

rate is coded into the machine tool in *inches per minute* (in./min) by the manufacturer.
- The feed rate used depends on the rigidity of the machine tool, the work setup, and the type of material machined.

The EIA standard feed rate code consists of the letter F plus five digits (three to the left and two to the right of the decimal point). The numbers to the left of the decimal point represent whole inches (or millimeters), and the numbers to the right of the decimal represent fractions of an inch (or millimeter).

- Generally when a feed rate in inches or millimeters per minute is programmed, it is given in whole numbers (to the left of the decimal). A feed rate per revolution is generally given in decimals (to the right of the decimal point).
- Feed can be programmed in inches (or millimeters) per minute (F25.5 would be a feed rate of 25 1/2 in./min).

Speed

The speed of a machine tool spindle generally means the number of revolutions the spindle makes in one minute of operation.

- The spindle speed rate (r/min) is generally governed by the work or cutter diameter, the type of cutting tool used, and the type of material being cut.
- Too fast a speed rate will cause the cutting tool to break down quickly, resulting in time wasted replacing or sharpening the cutting tool.
- Too slow a speed rate will result in the loss of valuable time, resulting in a higher cost for each part machined.

Therefore the speed rate is a very important factor that affects both the production rate and the life of the cutting tool. Various methods are used to set the spindle speed on CNC machine control units. The most common are by revolutions per minute (r/min), surface feet (or meters) per minute (sf/min or sm/min), and by the G96 function code which provides constant surface speed (CSS).

The EIA recommends that spindle speeds be programmed in revolutions per minute (r/min). The letter address S indicates spindle speed and may be followed by up to five digits. A spindle speed of 300 r/min would be programmed as S300; a speed of 2100 r/min would be S2100.

Spindle speed may be programmed in surface feet (or meters) per minute through the G96 preparatory function code. Some MCUs (especially on turning centers) have the capabilities of maintaining CSS at the point of the cutting tool. The proper value of the surface speed is programmed under an appropriate G function code, and as a diameter changes during a machining operation, the spindle speed will increase automatically, decrease, or remain unchanged.

MISCELLANEOUS FUNCTIONS

Miscellaneous CNC functions perform a variety of auxiliary commands, such as stopping the program, starting or stopping the spindle or feed, tool changes, coolant flow, etc., which control the machine tool. They are generally multi-character ON/OFF codes that select a function controlling the machine tool. Miscellaneous functions are used at the beginning or end of a cycle and are identified by the letter address M followed by a two-digit number.

Most miscellaneous codes such as M00, M01, M02, M06, or M26 are effective only in the specific block in which they are programmed. Most other miscellaneous codes do not have to be repeated in succeeding blocks.

Both preparatory and miscellaneous functions are generally classified as either *modal* or *nonmodal*. When *modal miscellaneous functions* such as M03 (spindle CW) and preparatory functions such as G81 (canned drill cycle) are programmed, they stay in effect in succeeding blocks until they are replaced by another function code. The modal function or code is changed or canceled as soon as a new preparatory function code is programmed.

All *nonmodal functions* such as M00, M01, M02, M06, etc., are valid or operational only in the block programmed. If they are needed in successive blocks, they must be programmed again.

The miscellaneous function codes in Fig. 4-6-13 have been supplied by the EIA and are listed in their standard EIA-274-D.

Code	Function
M00	Program stop
M01	Optional stop
M02	End of program (no rewind)
M03	Spindle start (forward CW)
M04	Spindle start (reverse CCW)
M05	Spindle stop
M06	Tool change
M07	Mist coolant on
M08	Flood coolant on
M09	Coolant off
M19	Spindle orientation
M30	End of program (return to top of memory)
M48	Override cancel release
M49	Override cancel
M98	Transfer to subprogram
M99	Transfer to main program (subprogram end)

Fig. 4-6-13 The most common EIA codes that control miscellaneous machine control functions.

PROGRAM FORMAT

The format on a CNC program refers to the sequence and arrangement of the coded information, which must conform to EIA standards. Each word address under the EIA standard should be listed in its proper place in the order of words. The following word addresses used in some controllers are used as an example to show their place in the order of words. There may be variations of this order of words with different MCU manufacturers.

Character Purpose

N	Sequence number for data block
G	Preparatory function
X	Amount of X axis travel (in. or mm)
Y	Amount of Y axis travel (in. or mm)
z	Amount of Z axis travel (in. or mm)
R	Clearance plane for fixed cycles (in. or mm)
I	Arc center coordinate parallel to X axis (in. or mm)

Character Purpose

J	Arc center coordinate parallel to Y axis (in. or mm)
K	Arc center coordinate parallel to Z axis (on Fanuc controls this is Q)
F	Feed rate or dwell
S	Spindle speed
T	Tool number
H	Tool length compensation
D	Cutter radius compensation
E	Fixture offset
L	Dwell time for fixed cycles
M	Miscellaneous function

SUMMARY

- The most common function codes used in CNC work are the preparatory (G) and the miscellaneous (M) codes.
- The structure of CNC programs consists of instructions containing technical, geometrical, and technological commands that include a block number, machine codes, word addresses, and technical information.
- A block of information consists of a sequence number, preparatory functions, miscellaneous functions, tool information, speeds and feeds, etc.
- Fixed or canned cycles are a preset combination of operations that can be recalled into a program whenever required to cause machine tool movement and spindle action to perform repetitive operations such as drilling, boring, turning, milling, etc.
- The location of any point on a flat surface can be accurately positioned by referring to where it is in relation to the X and Y axis.
- Coordinate information that includes axis movement, direction movement, and dimension movement should be programmed starting with the X first, Y second, and Z last.
- The R or work plane is set at either the highest surface of the part to be machined or at a specific height or distance above this surface.
- Miscellaneous CNC functions perform related operations such as starting or stopping the program, turning the spindle ON or OFF, coolant flow, changing cutting tools, etc.

KNOWLEDGE REVIEW

NC Functions

1. What two associations have established standards for CNC functions?
2. Name three types of function codes used for NC programming.

Block Format

3. Name the three types of commands that can be found in a CNC program.
4. What do the following word-address letters represent: G, F, S, M, T, X?

Sequence Numbers

5. Explain the importance of the sequence number.
6. How are the word address codes N, H, and O used?

7. What is the purpose of assigning sequence numbers in progression of 10?

Machine Motion Axes

8. How many motions does a single-spindle CNC drilling machine have?
9. Define each axis of a CNC drilling machine.
10. How can any point be positioned accurately on a flat plane?
11. Define "origin" or "zero point."
12. With a suitable illustration explain X and Y positive and negative dimensions.
13. In what sequence should axis motion be programmed if more than one axis is required?
14. Describe the Z axis and explain the difference between positive and negative movement.

15. What two motions occur when the Z axis is activated?
16. What does the word address letter R represent?
17. Define *work plane* and *gage height*.

Preparatory Functions

18. Name six operations that are activated by preparatory functions.

Fixed or Canned Cycles

19. Define fixed or canned cycles.
20. Why are canned cycles important in CNC work?
21. Explain what happens when a G81 drill cycle is programmed.
22. What types of codes are G82, G84, and G85?

Zero Suppression

23. Explain leading and following zeros.
24. What is meant by *zero suppression?*
25. When can both leading and trailing zeros be omitted in CNC programming?

Feeds and Speeds

26. Define feed.
27. What factors affect the feed rate used for machining?
28. What word address letter is used to program feed rate?
29. Define speed.
30. What factors affect the speed used during a machining operation?
31. Explain the effect of a speed rate which is: (a) Too slow (b) Too fast

Miscellaneous Functions

32. Name four auxiliary commands that miscellaneous functions perform.
33. Explain the difference between modal and nonmodal functions.
34. Define the following miscellaneous functions: (a) M00 (b) M02 *(c)* M03 (d) M06 (e) M30

Program Format

35. Define each word in the following block of information:
    N20 G83 X1.5 Y-0.5 R.100 Z-1.5 Q.250 M03

SECTION 5

Interpolation

INTERPOLATION

Interpolation, which is necessary for any type of programming, consists of generating data points between given coordinate axis positions. Within the Machine Control Unit (MCU), a device called an interpolator causes the drives to move simultaneously from the start to the end of the command. The interpolator is either an electronic hardware device for a NC system, or a software program for a CNC system. An interpolator provides two functions:

- It calculates individual axis velocities to drive the tool along the programmed path at the given feed rate.
- It generates thousands of intermediate coordinate points along the programmed path between the start point and the end point of the cut.

During positioning, all programmed axes move simultaneously at the specified feed rates until each axis has reached its destination. All drives start together, but without an interpolator individual destinations are reached successively according to the path traveled. However, an interpolator coordinates these axis motions in such a way that the programmed path is constantly maintained from the beginning to the end of the movement.

- Linear and circular interpolation are most commonly used in CNC programming applications:
- Linear interpolation is used for straight-line machining between two points.
- Circular interpolation is used for circles and arcs.
- Helical interpolation, used for threads and helical forms, is available on many CNC machines.
- Parabolic and cubic interpolation are used by industries that manufacture parts having complex shape such as aerospace parts, and dies for car bodies.

Interpolation is always performed under programmed feed rates.

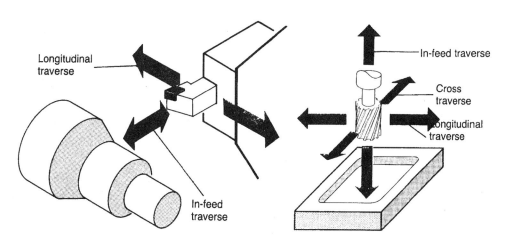

Fig. 5-1 A comparison of the linear interpolation axes for a CNC lathe and a mill. *(Hanser Publishers)*

The tools of a CNC machine can carry out certain axes movements depending on the machine type:

- On a lathe, the X axis controls the cross movement of the cutting tool and the Z axis controls the longitudinal movement of the cutting tool or carriage.

- On a vertical milling machine, the X axis controls the longitudinal movement of the machine table, the Y axis controls the cross movement of the table, and the Z axis controls the vertical movement of the spindle.

Linear Interpolations

Linear interpolation consists of any programmed points linked together by straight lines, whether the points are close together or far apart. The primary codes for linear interpolation (straight-line move) are G00 and G01. The G00 is a rapid traverse move used for positioning a cutting tool between two locations. The G01 command, with a programmed feed rate, is used for straight line cuts between two points parallel to the **X** and **Y** axis, Fig. 5-7-1A. Straight-line angular cuts are possible with the G01 command when the coordinate positions of the start and end point of the line are given along with a program feed rate, Fig. 5-7-1B.

OBJECTIVES

After completing this unit, you should be able to:

1. Prepare a program to position a cutting tool for drilling-type operations.
2. Write a program combining positioning and machining operations.
3. Prepare programs using the **X**, **Y**, and **Z** axes.
4. Write a program for cutting angular surfaces.

KEY TERMS

absolute positioning	gage height
angular programming	incremental programming
canned cycle	interpolation
continuous-path control	two-axes programming

Curves can be produced with linear interpolation by breaking them into short, straight-line segments. This method has limitations because a very large number of points would have to be programmed to describe a curve that would produce a contour shape. A contour programmed in linear interpolation requires the coordinate positions (**XY** positions in two-axis work) for the start and finish of each line or segment. Therefore, the end point of one line or segment becomes the start point for the next segment, and so on, throughout the entire program.

The accuracy of a circle or contour shape depends on the distance between two programmed points. If the programmed points (**XY**) are very close together, an accurate form will be produced.

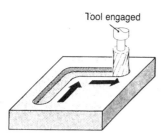

Fig. 5-7-1A G01 linear interpolation for straight line cuts parallel to the X and Y axis. *(Hanser Publishers)*

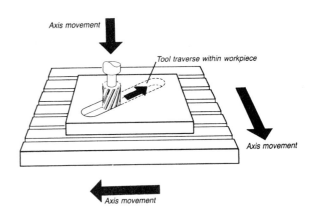

Fig. 5-7-1B G01 linear interpolation for straight line angular cuts. *(Hanser Publishers)*

To understand how a circle can be produced by linear interpolation, refer to Fig. 5-7-2. Figure 5-7-2B shows that when 8 connecting lines are machined, an octagon is produced, and Fig. 5-7-2C shows 16 connecting or chord lines. If the chords in each are examined, it should be apparent that the more program points there are, the more closely the form resembles a circle. If there were 120 chord lines, it would be difficult to see that the circle was made up of a series of short chord lines. Therefore, on a control unit that is capable only of linear interpolation, very accurate contours require very long programs because of the large number of points that need to be programmed.

REFERENCE POINTS

Every machine has a fixed reference point or selected location for each axis that is built in by

the manufacturer. Selected reference points can be established by the programmer before the programming process begins; they can be on the workpiece and on the cutting tool.

Machine reference

The machine zero or home position is the origin of the machine's coordinate system and is built in to the machine tool by the manufacturer. It is a fixed position and cannot be changed by the programmer or the machine operator.

All machine movements are measured from this reference or origin point. The MCU continually calculates the machine movements from the origin point. When a commanded distance or total of distances becomes more than the machine is capable of moving the MCU will display an

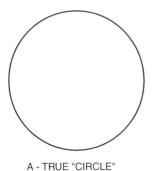

A - TRUE "CIRCLE"

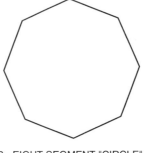

B - EIGHT SEGMENT "CIRCLE"

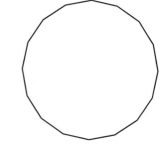

C - SIXTEEN SEGMENT "CIRCLE"

Fig. 5-7-2 Linear interpolation used to produce arcs and circles. (A) true circle (B) eight–segment circle C) sixteen segment circle. *(Allen Bradley Co.)*

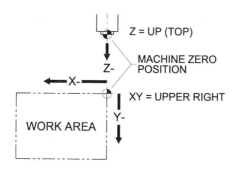

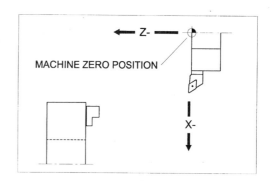

Fig. 5-7-3 Machine zero located at the upper right XY corner of the CNC vertical machining center table; Z zero machine tool change position. *(Peter Smid)*

Fig. 5-7-4 Machine zero position for a typical CNC lathe (rear tool type). *(Peter Smid)*

overtravel alarm stopping the machine from going past its limit of travel. On the MCU, the distance moved is displayed as the machine distance or relative distance.

On **vertical machining centers**, the machine origin for the **X** and **Y** axes is usually at the top right corner of the table. The **Z** axis is usually set at the tool change position, Fig. 5-7-3. The machine origin on the lathe is usually set at the top right side of the machine, Fig. 5-7-4. This is the maximum tool travel position for both the **X** and the **Z** axis in the positive direction or away from the workpiece. On all CNC machines using the standard coordinate system, the machine zero is located at the positive end of the travel for each axis.

WORKPIECE REFERENCE POINT

Somewhere within the machine motion limits, it is necessary to locate the workpiece to be machined. The workpiece must be mounted accurately and securely for any machining operation to produce consistent results and precision. It is very important to guarantee that each workpiece in the batch will be set the same way as the first workpiece. Once the setup is completed, a reference point on the workpiece must be selected. This point will be used in the program to establish the relationship between the machine reference point, the reference point of the cutting tool, and the drawing dimensions. See Figs. 5-7-5 and

5-7-6 for the common reference point of a CNC Mill and CNC Lathe.

Return to Machine Zero

It is the CNC operator's responsibility to physically reach the machine zero position (during setup) and to set the axes position in the control system. Normally, the power to the machine tool should never be turned off while the machine slides are right at or very close to the machine zero position; it makes the machine zero return more difficult later. A clearance of 1.0 inch (or 25 mm) and more is usually sufficient for each axis. A typical procedure to physically reach the machine zero position would follow these steps:

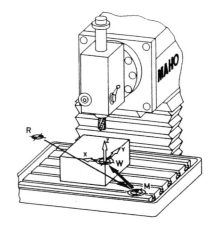

Fig. 5-7-5 The relationship between the part zero and the machine coordinate system for a vertical machining center. *(Deckel Maho, Inc.)*

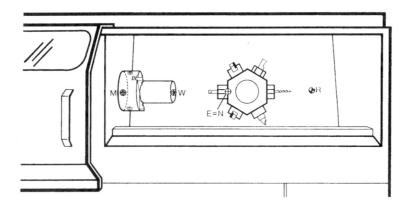

Fig. 5-7-6 The relationship between a CNC lathe part zero and the machine system coordinates. *(Peter Smid)*

1. Turn the power on (machine and control)
2. Select machine zero return mode
3. Select the first axis to move (usually **Z** axis)
4. Repeat for all the other axes
5. Check the lighted 'in position' indicators
6. Check the POSITION display
7. Set display to zero if necessary

Usually the first axis selected is the **Z** axis for CNC machining centers, and the **X** axis for CNC turning centers because they will be moving away from the work into the clear area. When the axis has reached the machine zero position, a small indicator light on the control panel turns ON to confirm that the axis actually reached the machine zero position. The machine is now at its reference position, at the machine zero or at the machine reference point whichever term will be used. The indicator light is the confirmation for each axis. Although the machine tool is ready for use, a good CNC operator will go one step further. On the display screen, marked POSITION, the actual relative position should be set to zero readout for each axis, as a standard practice.

PROGRAM ZERO - MACHINING CENTER

It is the CNC programmer who determines the setup method for a given job, perhaps in cooperation with the CNC operator. It is also the CNC programmer who selects the program zero posi-

tion for each program. The process of selecting the program zero starts when the drawing is evaluated. The following two steps have to be completed first:

Step 1.—Study how the drawing is dimensioned and determine which dimensions are critical and which are not.

Step 2.—Decide on the method of setting up and holding the workpiece. Once this has been done, select the program zero that is most suitable.

— Common setup methods used are vises, chucks, subplates, clamping to the machine table, and hundreds of special fixtures, Fig. 5-7-7A,B,C.

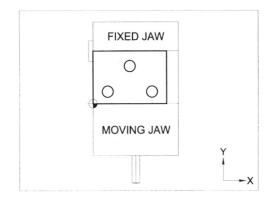

Fig. 5-7-7A Part in a vise with the XY zero located on the movable vise jaw. (not recommended) *(Peter Smid)*

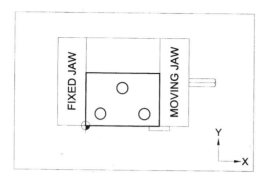

Fig. 5-7-7B   Part mounted in a machine vise using the solid jaw as a reference location. *(Peter Smid)*

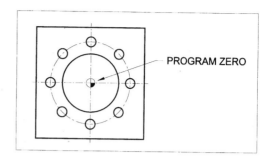

Fig. 5-7-8   A program zero for circular objects is usually the center of the part or circle. *(Peter Smid)*

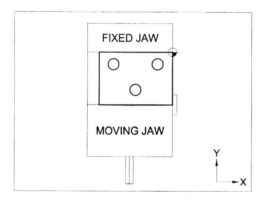

Fig. 5-7-7C   Part mounted in a machine vise using the fixed jaw as a reference location. *(Peter Smid)*

To select a program zero, all three axes must be considered. The workpiece reference point is commonly known as the program zero or part zero. It can be located at any point on the part that is most suitable for the programmer or the machining operation to be performed.

Program Zero Selection

The selection of the program zero, often influences the efficiency of the workpiece setup and the machining operation, Fig. 5-7-8. When programming circular workpieces it may be easier to program and set up the part using the center as its program zero.

In theory, the program zero point may be located anywhere where it is most convenient. Two factors should be kept in mind when select-

ing the program zero, machining accuracy, and convenience of setup and operation.

- **Machining accuracy** (the most important)
  - All the workpieces in the batch must be machined to the drawing specifications. Accuracy is the difference between the expected value and the measured value. Every piece in the batch must be the same as the previous ones.
- **Operating and setup convenience**
  - Convenience should only be considered after the machining accuracy is assured.
  - An experienced CNC programmer will always think of the effect the program has in the machine shop.

Defining the program zero that is difficult to set on the machine or difficult to check, can be inconvenient and may reduce the productivity of CNC operator.

MACHINE PROGRAM ZERO

The typical methods and common considerations of program zero selection for the vertical CNC machining centers and the CNC turning centers are considered individually.

Program Zero—Lathes

On CNC lathes, the program zero selection has only two axes to consider the **X** axis and the **Z** axis. Because a lathe produces cylindrical parts, the program zero for the **X** axis must always be on the spindle centerline. There are three popular methods of selecting the **Z** axis program zero for a typical workpiece on a CNC lathe, Fig. 5-7-9:

> Chuckface ... Main face of the chuck
> Jawface ... Locating face of the jaws
> Partface ... Right end of the finished workpiece

The most popular method on modern CNC lathes is to select the program zero on the end face of the workpiece. It is recommended that this be a finished face or if not, it should be machined first before the part zero is set.

The benefits of the end face program zero are that the many Z dimensions specified on the drawing can be transferred directly into the CNC program. A great deal depends on the way the engineering drawing is dimensioned, but the CNC programmer usually benefits anyway. The additional benefit, probably the best liked, is that a negative **Z** value of a tool motion indicates the work area, and a positive **Z** value is the clear area. When writing the program, it is easy to forget to assign the minus sign to the **Z** axis cutting motions that will position the cutting tool in the area away from the work. The examples in this book use program zero at the end finished face, unless otherwise noted.

TOOL REFERENCE POINT

The reference point of the cutting tool differs between the tools used for milling and those used for turning. In milling-related operations, the reference point of the tool is usually the intersection of the tool centerline and the lowest positioned cutting tip (edge). In turning, the most common tool reference point is the imaginary tool point of the cutting insert.

For tools such as drills, reamers, taps, standard end mills, and most of other point-to-point tools, regardless whether they are used on milling or turning machines, there is no confusion about where the reference point of the tool is located. It is always at the most extreme tip as measured along the **Z** axis, Fig. 5-7-10.

LINEAR PROGRAMMING EXAMPLES

Linear programming may be performed in the G90 absolute mode or the G91 incremental mode. In the absolute mode, all locations (measurements) are taken from the same program zero. In the incremental mode, each program location (measurement) is taken from the previous point. Figure 5-7-11 shows a milling exercise that will be programmed in both the absolute and incremental mode to show the differences between the two programming systems. Each programmed step will be explained so that the student becomes familiar with the codes and the programming sequences.

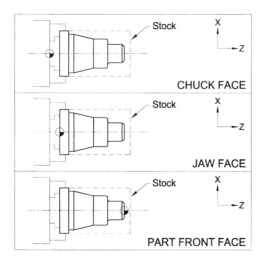

Fig. 5-7-9   Three common program zero positions used for CNC lathes. *(Peter Smid)*

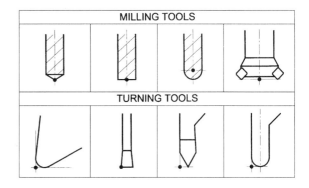

Fig. 5-7-10   Tool reference position for various cutting tools. *(Peter Smid)*

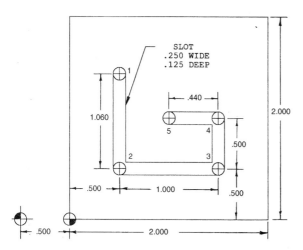

Fig. 5-7-11 An example of linear interpolation machining along the X and Y axes. *(Kelmar Associates)*

CODES NEW TO THIS UNIT

G00	Rapid positioning
G01	Linear interpolation (straight-line move)
G20	Inch input mode
G28	Machine zero return (reference point 1)
G43	Tool length compensation (offset) - positive
G54	Work offset position register
G90	Absolute dimensioning mode
G91	Incremental dimensioning mode
G96	Constant surface speed mode (lathe)

* * * * **

M01	Optional program stop
M03	Spindle rotation (clockwise)
M05	Spindle stop
M08	Coolant ON
M09	Coolant OFF
M30	Program end (always with reset and rewind)
M41	Low gear selection

* * * * * *

F	Feed rate in in./ or mm/rev.
H01	Tool length offset 1
I	Taper amount and direction/side (taper turning)
O	Letter O precedes the program number
S	Spindle speed in r/min.
T01	Tool number 1
U	X axis incremental move (turning)
W	Z axis absolute move (turning)

Absolute Programming

O6788		- Program number (Use letter O as the first digit)
N010	G90 G20	
	G90	- Absolute programming
	G20	- Inch input mode
N020	G00 G54 X.500 Y1.560 S2000 M03	
	G00	- Rapid traverse mode
	G54	- The work offset
	X.500) Y1.560)	Location of start point
	S2000	Spindle speed 2000 r/min.
	M03	- Spindle ON clockwise
N030	G43 Z.100 H01 M08	
	G43	- Tool length offset command (positive)
	Z.100	- Tool .100 above work surface
	H01	- Tool length offset
	M08	- Coolant ON
N040	G01 Z-.125 F5.0	
	G01	- Linear interpolation mode
	Z-.125	Depth of cut in absolute
	F5.0	- Feed rate 5 in/min.
N050	Y.500	Groove cut from point 1 to 2
N060	X1.500	Groove cut from point 2 to 3
N070	Y1.000	Groove cut from point 3 to 4
N080	X1.060	Groove cut from point 4 to 5
N090	G00 Z.100 M09	
	Z.100	Tool rapids to .100 above work surface
	M09	- Coolant OFF
N100	G28 Z.100 M05	
	G28	- Machine zero return
	M05	- Spindle stop

N110   G28   X1.060   Y1.000
N120   M30   End of program (rewind code)
%

Incremental Programming

The same part used in Fig. 5-7-11 will be programmed in the incremental mode where the end of a previous point will be the start of the next point.

O6789
N010   G90   G20
N020   G00   G54   X.500   Y1.560   S2000   M03
       (It is common practice to locate the tool at the start point in the absolute mode)
N030   G43   Z.100   H01   M08
N040   G91   G01   Z-.225   F5.0
       G91   -   Incremental mode
       G01   -   Linear interpolation
       Z-.225   Tool feeds to .125 depth (.225-.100)
N050   Y-1.060   Tool moves from point 1 to 2
N060   X1.000   Tool moves from point 2 to 3
N070   Y.500   Tool moves from point 3 to 4
N080   X-.440   Tool moves from point 4 to 5
N090   G00 Z.225   Tool rapids to .100 above
  M09   work surface
N100   G28   Z0   M05
N110   G28   X0   Y0
N120   M30
%

Lathe Programming

Figure 5-7-12A shows a turning exercise that will be programmed in both the absolute and incremental mode to show the differences between the two programming systems. In CNC lathe programming using the **XZ** prefixes sets the program automatically in the absolute mode; **UW** is used for the incremental mode, Fig. 5-7-12B. Each programmed step will be explained so that the student becomes familiar with the codes and the programming sequences.

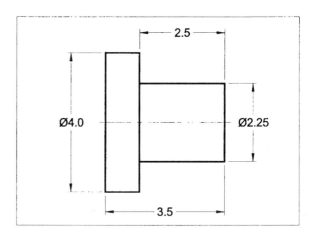

Fig. 5-7-12A A CNC lathe turning exercise for programming and machining parallel diameters. *(Peter Smid)*

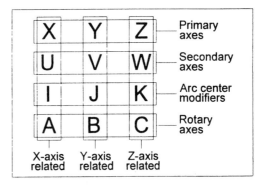

Fig. 5-7-12B The relationship of the primary and secondary machine axes. *(Peter Smid)*

Absolute Programming (Straight Turning)

O1001
N010   G20
N020   T0100   M41
       T0100   -   Tool number
       M41   -   Low gear selection
N030   G96   S450   M03
       G96   -   Constant surface speed mode
N040   G00   X4.200   Z.100   T0101 M08
                              (Start point)
N050   G90   X3.662   Z-2.495   F.010
                              (Pass 1)
N060   X3.324   (Pass 2)

N070   X2.986   (Pass 3)
N080   X2.648   (Pass 4)
N090   X2.310   (Pass 5)
N100   G00 X8.0 Z3.0 T0100 M09
N110   M01   (end of roughing)
       M01   -   Optional program stop
%

Incremental Programming (Straight Turning)

The same part used in Fig. 5-7-12A will be programmed in the incremental (using the **UW** prefix) mode where the end of a previous point is the start of the next point.

O1002
N010   G20
N020   T0100 M41
N030   G96 S400 M03
N040   G00 X4.2 Z.100 T0101 M08
N050   G90 U-.538 W-2.595 F.012
N060   U-.338
N070   U-.338
N080   U-.338
N090   U-.338
N100   G00 X8.0 Z3.0 T0100 M09
N110   M30
%

TAPER TURNING

Tapers are used in the mechanical trades to provide an accurate method of quickly and accurately aligning machine parts, and holding cutting tool accessories such as adapters, milling cutters, etc. Taper turning on a CNC lathe is similar to machining angular sections on a CNC mill. The information that must be provided when programming tapers are the coordinate dimensions of the start point diameter, the end point diameter, and a feed rate.

Taper Turning (small angle tapers)

Machining small-angle tapers, or those that can be cut in one programmed pass, can be programmed in a manner very similar to straight turning. The program line for the taper section

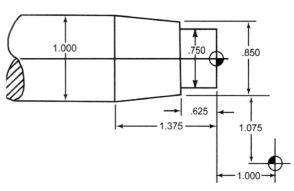

Fig. 5-7-13 Small–angle tapers can be programmed in one block and machined in one pass. *(Kelmar Associates)*

must contain the small diameter dimension, the large diameter dimension, and the length of the tapered section. Absolute programming will be used to program the part shown in Fig. 5-7-13.

Programming the tapered part

O1003
N010   G20
N020   G00 G54
N030   T0100 M41
N040   G96 S400 M03
N050   G00 X-.100 Z.010 M08
N060   G01 X0 Z0 F.010
N070   X.750
N080   Z-.625
N080   X.850
N090   X1.000 Z-1.375
N100   G00 X1.000 Z1.500 M09
N110   M30
%

Taper Turning (steep tapers)

The part shown in Fig. 5-7-14 has a steep taper over a short distance that will require more than one cutting pass to machine the tapered section. When machining steep tapers that require more than one block of program information because of the amount of material to be removed, it is necessary to add the **I** value to the cycle call. This is called the **single radius value** and it indicates the taper amount and its direction per side relative to the **X** axis. The **I**

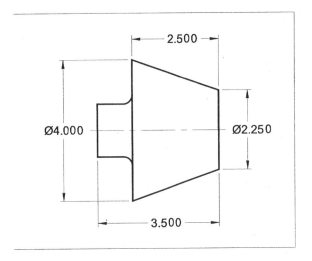

Fig. 5-7-14 Steep tapers can be programmed in one block using the I value to allow multiple passes in the machining cycle. *(Peter Smid)*

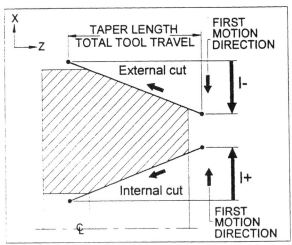

Fig. 5-7-15A The I value used for G90 turning cycle – external and internal. *(Peter Smid)*

value is calculated as a single distance per single side (radius value) for the total traveled distance.

Two rules apply when using the **I** value for the total traveled distance.

- If the direction of the first tool motion is **X** negative (**X–**), the **I** value is **I–,** Fig, 5-7-15A (top).
- If the direction of the first tool movement is **X** positive (**X+**), the **I** value is **I+,** Fig. 5-7-15A (bottom).

To calculate the **I** value for the part in Fig. 5-7-14, add .100 in. to each end of the taper length to allow for tool clearance. This increases the tool travel length to 2.700 in. along the Z axis and this is the length used to calculate the **I** value for programming the steep taper, Fig. 5-7-15B. Using trigonometry, this value would be: **I = 2.700 × tan a** or .945 in., Fig. 5-7-16.

The steep-taper part shown in Fig. 5-7-14 will be used to program a CNC lathe in both the absolute and incremental mode.

Taper Turning—steep taper (absolute)

O1003
N010 T0100 M41

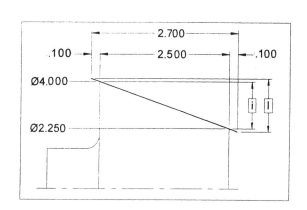

Fig. 5-7-15B Known and unknown values for taper turning – Program O1003. Value I is known, value I must be calculated. *(Peter Smid)*

N020	G96 S400 M03	
N030	G00 X4.2 Z.100 T0101 M08	(Start)
N040	G90 X3.662 Z-2.495 I-.945 F.012	(1)
N050	X3.324	(2)
N060	X2.986	(3)
N070	X2.648	(4)
N080	X2.31	(5)
N090	G00 X8.0 Z3.0 T0100 M09	
		(clear POS)
N100	M01	(End of roughing)
N110	M30	
%		

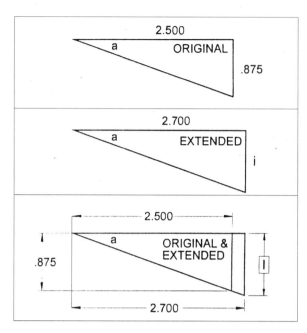

Fig. 5-7-16 Information required for calculating the I distance required for programming steep tapers. *(Peter Smid)*

Taper Turning - steep taper (Incremental)

```
O1004
N010   G20
N020   T0100   M41
N030   G96   S400   M03
N040   G00   X4.2   Z.100   T0101   M08
N050   G90   U-.538   W-2.595   I-.945   F.012
N060   U-.338
N070   U-.338
N080   U-.338
N090   U-.338
N100   G00   X8.0   Z3.0   T0100   M09
N110   M30
%
```

LINEAR INTERPOLATION (ANGLES)

Linear interpolation involves moving the cutting tool from one position to another in a straight line. With this type of programming, any straight-line section can be machined, including all tapers or angular surfaces. When linear moves are programmed, the coordinates (**XY** axes) for the beginning and end of each line must be given. Linear interpolation may also be used to simulate arcs and circles, but it is not often used for this purpose because of the large number of coordinate locations that must be programmed to keep each segment move as short as possible. The smaller each segment is, the smoother the arc or circle, Fig. 5-7-17.

Angular Programming

Whenever movement is required along two axes (**X** and **Y**), the axes move simultaneously along a vector path, Fig. 5-7-18. The rate of travel along the vector path is set automatically by the machine control unit (MCU) so that it is equal to the programmed feed rate. An angle is a straight line connecting a start point and an end point, so

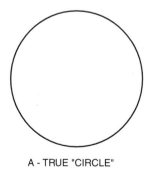
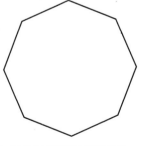
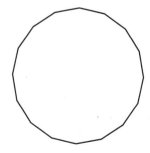

A - TRUE "CIRCLE"   B - EIGHT SEGMENT "CIRCLE"   C - SIXTEEN SEGMENT "CIRCLE"

Fig. 5-7-17 Linear interpolation used to produce arcs and circles. (A) true circle (B) eight–segment circle C) sixteen segment circle. *(Allen Bradley Co.)*

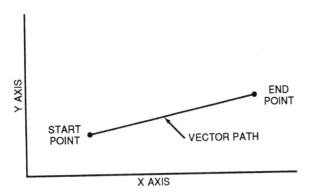

Fig. 5-7-18 The locations of the start point and the end point, and a feed rate are required to program angular cuts. *(Kelmar Associates)*

linear interpolation can be used to produce angles and tapers regardless of length. Once the two points have been programmed, along with a vector feed rate, the MCU holds this information in its memory until the end point of the line is reached. Therefore the function of interpolation is to store information and constantly compare and correct the machine axis motions to keep a straight-line movement between the start point and end point coordinates at a specified vector feed rate.

The sample workpiece shown in Fig. 5-7-19 requires that a 30° angular slot .250 in. wide by .125 in. deep be cut. Before this operation is programmed it is first necessary to calculate the coordinate positions (**XY**) of the start point and

end point of the angular slot. If this cut is to be programmed in the incremental mode, the calculations are as follows:

Start point   X1.500 Y.250
End point   X2.000 Y1.155
Feed rate   10 in./min

The part shown in Fig. 5-7-20 combines linear and angular interpolation to program and machine a slot that requires straight line and angular cutting. The slot is .250 in. wide and .125 in. deep and should be cut in one pass of the cutting tool with the starting point at the top left corner of the groove.

Angular Programming (Mill)—absolute

01114
N010 G20
        G20   -   Inch data input
N020 G00 G54 X.565 Y1.500 S2000 M03
                                      (Point 1)
        G54   -   Work offset
N030 G43 Z.100 H01 M08
        G43   -   Tool length offset
                           command (positive)
        H01   -   Tool length offset
N040 G01 Z-.125 F5.0
N050 X1.505 Y.500   (Point 2)
N060 Y1.500   (Point 3)
N070 X.565   (Point 1)

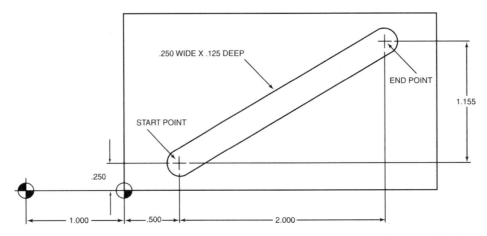

Fig. 5-7-19 A workpiece that requires an angular slot to be cut. *(Kelmar Associates)*

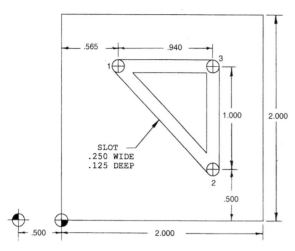

Fig. 5-7-20 A triangular slot must be programmed and machined using both the absolute and incremental programming modes. *(Kelmar Associates)*

N080 G00 Z.100 M09
      M09   -   Coolant OFF
N090 G28 Z.100 M05
      M05   -   Spindle stop
N100 G28 X.565 Y1.500
N110 M30
%

Angular Programming (Mill)—Incremental

O1113
N010 G20 G90
N020 G00 G54 X.565 Y1.500 S2000 M03
N030 G43 Z.100 H01 M08
N040 G91 G01 Z-.225 F5.0
N050 X.940 Y-1.000
N060 Y1.000
N070 X-.940
N080 G00 Z.225 M09
N090 G28 Z0 M05
N100 G28 X0 Y0
N110 M30
%

SUMMARY

- The interpolator is a vital part of the MCU, allowing the simultaneous movement of two or more axes.
- The coordinated movement of these axis allows the machine tool to move the cutter or the workpiece in a constant tool path to generate:
    - Linear interpolation - straight line and angular moves.
    - Circular interpolation - arcs and circular moves.
    - Helical interpolation - threads and helical forms.
    - Parabolic and cubic interpolation - for complex shapes.
- Reference points provide a fixed or selected position which is either established by the machine tool manufacturer, by the programmer or, by the operator.
- Reference points provide a machine home position, a workpiece position, a program zero position and a tool reference point:
    - **Machine home position** - is the location in each axis from which the machine tool calculates its movements.
    - **Workpiece reference position** - is selected by the programmer or operator to set up the workpiece for machining.
    - **Program zero** - the position on the workpiece used as a reference point to write the CNC program.
    - **Tool reference point** - the contact point on the cutting tool used by the programmer for programming and setup:
        - On milling cutters the extreme tip of the tool as measured along the **Z** axis; for turning tools, the imaginary tool point of the cutting insert.

KNOWLEDGE REVIEW

1. Name two types of reference points?
2. What is the machine zero or home position?
3. Who establishes the machine home position?
4. What happens if a commanded distance is larger than the machine is able to move?
5. Where is the **X**, **Y**, and **Z** machine origin usually located on a machining center?
6. Briefly explain the workpiece reference point.
7. Who is responsible for setting the machine zero?
8. What is the first axis selected to zero return on the mill and why?
9. How does the operator know when the machine has reached this location?
10. Briefly describe the purpose of the program zero position.
11. Who selects the part program zero position?
12. What are two considerations to selecting the part program zero?
13. What are four common methods of holding a workpiece to be machined?
14. When programming circular workpieces where is an easy program zero position to use?
15. Where is the program zero usually selected for a CNC lathe part?
16. What are the three most popular **Z** zero program axis positions used on a CNC lathe?
17. What is the **Z** axis tool reference point for cutting tools?

Unit 8
Circular Interpolation

Circular interpolation was developed to overcome the difficulty in programming arcs and circles. It allows a programmer to make the cutting tool follow any circular path ranging from a small arc segment to a full 360° circle for machining arcs or full circles, outside and inside radii, circular pockets, radial recesses and grooves, helical cutting, etc. On some MCUs, the coordinate locations of the start point and end point of the arc, the radius of the circle, the coordinate location of the circle center, and the direction of cutter travel must be programmed, Fig. 5-8-1. The circular interpolation of the MCU automatically breaks up the arc into very small linear moves, generally .0001 or .0002 in. (0.0025 or 0.005 mm) each, to describe the circular path. The MCU then generates the controlling signals to move the cutting tool to produce the desired arc or circle. If the same arc or circle were to be programmed by linear interpolation, hundreds or even thousands of coordinates, each defining a span, would have to be programmed.

OBJECTIVES

After completing this unit you should be able to:

1. Write a program to machine partial arcs and contour forms.
2. Know the procedure for programming a circle using the quadrant method.
3. Program a circle using the full circle (360°) method.

KEY TERMS

arc

arc center modifiers

arc end point

arc start point

circular interpolation

quadrant points

radius programming

tool path

CIRCLE ELEMENTS

It is common in CNC work to machine arcs and circles so it is wise to understand the elements of a circle. A common definition for a circle found in a dictionary is as follows:

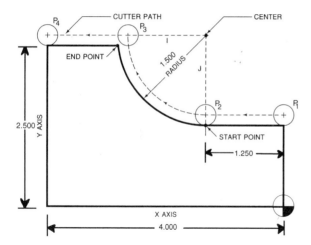

Fig. 5-8-1 Two–dimensional circular interpolation requires a start point, end point, coordinate location of the arc center, and the direction of cutter travel to be programmed. *(Kelmar Associates)*

> *A circle is a plane figure bounded by a curved line where every point on its circumference is equidistant from its center point.*

The main elements of a circle are: diameter, radius, quadrants, and quadrant points, Fig. 5-8-2.

Radius and Diameter

The two most important parts of a circle used in circular programming are the radius and the diameter.

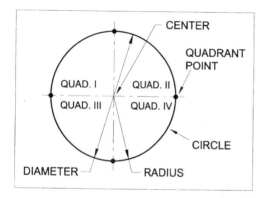

Fig. 5-8-2 A circle contains four basic elements that are important in CNC programming. *(Peter Smid)*

> *The radius of a circle is the line from the circle center to any point on the circle circumference.*

> *The diameter of a circle is the line through the center of the circle with both end points touching the circle circumference.*

Quadrants

A circle is divided into four quadrants or major sections, Fig. 5-8-2. A circle is programmed in all four quadrants and an arc may be programmed within one or more quadrants. Arc modifiers I J K (described later) are used to define a unique arc radius.

The **R** address is fairly common on most machine control systems for arc programming. When using the R address in programming, it is important to provide the start and end points of the arc because the computer uses this information to find the center point of the arc.

- Arcs with an angular difference of 180° of less, measured between the start and end points, use a **positive** value of the **R** address.
- Arcs where the angular difference is **more that 180°,** but not 360°, use a negative address **R.**

Quadrant Points

As seen in Fig. 5-8-2, quadrant points are the four points at the ends of the vertical and horizontal lines. The point where any of these lines intersect the circle circumference is called the **Quadrant Point** or **Cardinal Point**, Fig. 5-8-3. These points become very important when programming partial circles.

PROGRAMMING ARCS

Radius and diameter dimensions are commonly used on technical drawings to accurately describe and dimension arcs and circles. They are also commonly used in turning and machining centers to describe tool and part radii, cutter offset dimensions, and center-point locations.

Degrees	Compass Direction	Watch Direction	Bound by quadrants
0	East	3 o'clock	IV and I
90	North	12 o'clock	I and II
180	West	9 o'clock	II and III
270	South	6 o'clock	III and IV

Fig. 5-8-3 The location and relationship between the four quadrants of a circle. *(Peter Smid)*

There are various control units on the market. Most of these units can generate a full circle (radius programming) at one time but some of the older models generate only one quadrant of a circle (center-point programming) at a time. Some models of MCUs are limited to circular interpolation in a two-axis plane at a time, such as XY, XZ, or YZ axes, while others can interpolate circular movements for three axes for machining helixes. Circular interpolation can also be used to generate second- and third-degree curves and free-form shapes that can be closely described with a series of arcs or circles. The programming for controlling a tool path to produce an arc is almost the same as programming along a straight line.

When circular interpolation is programmed, three pieces of information are necessary:

1. The direction of the cutter travel (preparatory function)
2. Arc start and end points (X Y coordinates)
3. The center point of the arc and the radius value (**R or IJK**)

The *direction of cutter travel is* defined by the standard EIA preparatory function codes for circular interpolation. G02 is circular interpolation in a clockwise direction (CW). G03 represents circular interpolation in a counterclockwise direction (CCW), Fig. 5-8-4. These codes must be programmed in the block of information where circular interpolation starts, and they remain effective (modal) until a new preparatory (G) code is programmed.

• The *start point of arc* is usually the end point of a linear line or the end point of a previous arc.

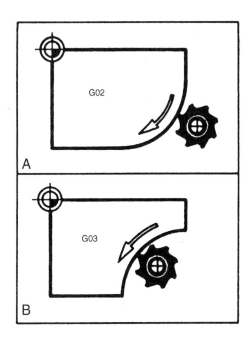

Fig. 5-8-4 The G02 command (Fig. A) causes the tool to move in a clockwise direction and the G03 command (Fig. B) causes the tool to move in a counterclockwise direction. *(Deckel Maho, Inc.)*

The start point sets the position of the cutting tool for machining the arc and is generally given as XY and/or Z coordinate dimensions.

• The *center point of the arc* (XY and/or Z coordinates) is the center of the circle or arc and is described by I (X coordinate value), J (Y coordinate value), and K (Z coordinate value), Figs. 5-8-5A and 5B. Generally the I, J, and K words are incremental values regardless of whether they have been programmed in the absolute or incremental mode. The I and J (Mill) and I and K (Lathe) coordinate values are always taken from the start point of each arc or 90° segment to the center point of the radius, Figs. 5-8-5A and 5B.

• The *end point of the arc* (XY and/or Z coordinates) is the last point where the cutter path center line completes the circular path. Whenever an arc uses more than one 90° quadrant, the point where it crosses into the next quadrant must be programmed as the end point. The MCU assumes that this is also the start point for the next quadrant; therefore it is only nec-

THE CIRCLE CENTER COORDINATES I AND J

When the programming of circles and arcs with the address R is not possible, the circle center has to be programmed. This is done with the addresses I and J:

I is a coordinate parallel to the   X axis
J is a coordinate parallel to the   Y axis.

G 17

Vertical Machining Center

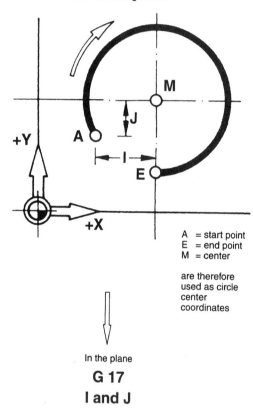

A = start point
E = end point
M = center

are therefore used as circle center coordinates

In the plane
G 17
I and J

Fig. 5-8-5A   The information required to program arcs and circles. *(Deckel Maho, Inc.)*

essary to program the end point for the next quadrant.

ARC CENTER INFORMATION

To eliminate problems and produce accurate arcs while using circular interpolation, it is important that the block line in the program contain the following information:

• Arc cutting direction (CW or CCW)

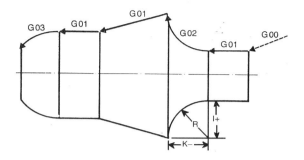

Fig. 5-8-5B   The I and J arc center modifiers used for arc turning operations. *(Deckel Maho, Inc.)*

• Arc start and end point coordinate positions
• Arc center and radius value

In addition to this information the feedrate must also be included in the program line to move the cutting tool along the arc path.

Arc Cutting Direction

A cutting tool may be moved along an arc in a clockwise (CW) or counterclockwise (CCW) direction. This cutting direction information should be given in the first statement of the block. The two preparatory motion commands for arcs are G02 (clockwise) and G03 (counterclockwise). They are modal commands that stay in effect until replaced by another command from the same group, usually another motion command such as G00 or G01.

When the G02 or G03 command is activated by the program, any previous motion command (usually G00, G01, or cycle command) in the program is automatically cancelled. All circular toolpath motions must have a cutting feedrate in effect and it must be programmed before or within the cutting motion block.

Arc Center and Radius

The radius of the arc can be designated with the address **R** or with arc center modifiers **I, J, and K**, Fig. 5-8-6. The **R** address allows the arc radius to be programmed directly, and the **IJK** arc center modifiers are used to define the position of the arc center. Most modern control systems support the **R** address input, older controls require the arc

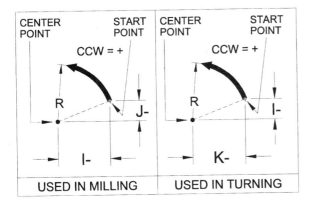

Fig. 5-8-6 Lathe and Mill arc center points and start points. *(Peter Smid)*

center modifiers. The basic programming format will vary only slightly between the milling and the turning systems, particularly for the **R** address version:

G02 X.. Y.. R..   Milling program—CW
G02 X.. Z.. R..   Turning program—CW
G03 X.. Y.. R..   Milling program—CCW
G03 X.. Z.. R..   Turning program—CCW

The arc center location and the arc radius are very important along with a circular interpolation mode to produce an arc accurately. Keep in mind that numerical control means control of the cutting toolpath by numbers and there are a great many arc radii that will fit between the programmed start and end points and still maintain the cutting direction. The control system needs additional information that identifies the programmed arc with a unique radius.

This unique radius is achieved by programming the **R** address for the direct radius input, or using the **IJK** arc center modifiers. Address **R** is the actual radius of the toolpath (usually the radius taken from the part drawing).

Arc Center Modifiers

Arc center modifiers (**IJK**) are used to accurately describe the distance from the start point of an arc to the center point of the arc. This distance is almost always measured as an incremental distance between two points. Some control systems

use the absolute designation to define an arc center. The arc center is then programmed as an absolute value from the program zero, not from the arc center. It is important to know how the control systems of the machine tool being used handles these situations.

Fig. 5-8-7 shows the signs of arc modifiers **I** and **J** in all possible orientations. In different planes, different pairs of modifiers are used, but the logic of their usage remains the same. Arc modifiers **I, J, and K** are used according to the following definitions:

> Arc center modifer **I** is the distance, with specified direction, measured from the start point of the arc to the center of the arc, parallel to the X axis.

> Arc center modifier **J** is the distance, with specified direction, measured from the start point of the arc to the center of the arc, parallel to the Y axis.

> Arc center modifier **K** is the distance, with specified direction, measured from the start point of the arc to the center of the arc, parallel to the Z axis.

Programming an Arc (Center Point Method)

The two most common methods of programming an arc are by center-point method and by radius method. Some MCUs will generate an arc if it is defined by the arc center point (XY coordinates) and the end point of the arc. Other MCUs will generate an arc if it is defined by the radius and the end point of the arc. Before an arc or circle is programmed it is necessary to determine what information is needed for the particular MCU being used.

There are three questions that must be answered before an arc is programmed. Place a pencil point at the start point of the arc and answer the following questions before the pencil point is moved.

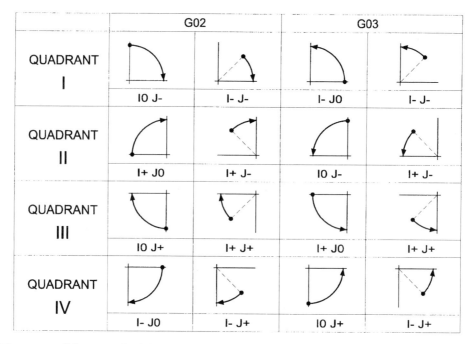

	G02		G03	
QUADRANT I	I0 J-	I- J-	I- J0	I- J-
QUADRANT II	I+ J0	I+ J-	I0 J-	I+ J-
QUADRANT III	I0 J+	I+ J+	I+ J0	I+ J+
QUADRANT IV	I- J0	I- J+	I0 J+	I- J+

Fig. 5-8-7 The arc modifiers I and J (also known as vectors) and their sign designations in various quadrants. *(Peter Smid)*

1. Which way? Clockwise (G02) or counter-clockwise (G03) direction from the start point of the arc.
2. Where to? The X and Y coordinates of the end point of the arc.
3. How far? The I and J values from the start point of the arc to the center of the circle.

The information required for center-point programming is as follows:

1. G-code. G02 circular interpolation clockwise, G03 circular interpolation counterclockwise.
2. End point. The X and Y coordinates of the end point of the arc.
3. Center point. The coordinates of the center point of the arc. The letters I (X axis) and J (Y axis) are used to define the point.

The information required for radius programming is as follows:

1. G-code. G02 circular interpolation clockwise, G03 for circular interpolation counterclockwise.

2. End point. The X and Y coordinates of the end point of the arc.
3. Radius. The radius of the arc preceded by the letter address R.

CODES NEW TO THIS UNIT

G02	Circular interpolation (clockwise)
G03	Circular interpolation (counterclockwise)
I	Arc Center Modifier - Distance from start of arc to center of arc
J	Arc Center Modifier - Distance from start of arc to center of arc
K	Arc Center Modifier - Distance from start of arc to centr of arc
R	Direct radius address

To Program an Arc (Center Point)

For the two arcs shown in Fig. 5-8-8 to be programmed, the start point and the end point for each arc must be programmed. If the incremental

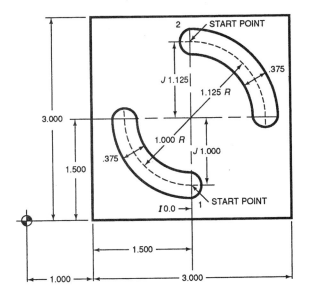

Fig. 5-8-8 The I and J distances are taken from the start point of the arc to the center point of the radius. (*Kelmar Associates*)

mode is used, the programming for the two arcs .375 in. wide by .250 in. deep would be as follows:

%		Program start code
O1234		Program number
N010	G20	
N030	G90 G00 G54 G43 X2.500 Y.500 Z.100 S550 M03	
	G90	Absolute positioning mode
	G00	Rapid to position #1 (X2.500 Y.500).
	G54	Work offset position register.
	G43	Tool length compensation (positive).
	Z.100	Spindle rapids down to .100 above the work surface.
	S550	Spindle speed.
	M03	Spindle ON clockwise.
N040	G91 G01 Z-.250 F2.0	
	G91	Incremental positioning made.
	Z-.250	Cutter feeds into the workpiece .250.

	F2.0	2 in. feed rate.
N050	G02 X-1.000 Y1.000 1.0 J1.000 F10.0	
	G02	Circular interpolation clockwise.
	X-1.000)	Locates the coordinate position of the end point
	)	of the arc.
	Y1.000)	
	I0.0)	The center of the radius is located from the
	)	start point of the arc.
	J1.000)	
N060	G00 Z.250	
	The cutter rapids to .250 above the work surface.	
N070	X1.000 Y1.125	
	Rapid to position #2.	
N080	G01 Z-.250 F2.0	
	The cutter feeds to .250 depth.	
N90	G02 X1.125 Y-1.125 I0.0 J-1.125 F10	
	Same as sequence N050.	
N100	G00 Z.100	
	Same as sequence N060.	
N110	X-3.625 Y-1.500 M06	
	Rapid back to XY zero (the start position).	
	M06	stops the spindle and retracts to full retract position.
N120	M30	
	M30	Rewinds the program for the next part.
%		Program end code.

To Program a Circle (Center Point)

A full circle consists of four 90° arcs, and since some MCUs generate only one quadrant at a time, each arc must be programmed. The end point of one arc automatically becomes the start point of the next arc, so a full circle requires the *first start point* and *four end points* to be programmed. The program for the .375 in. wide by .250 in. deep circular groove shown in Fig. 5-8-9 is as follows:

O2345		Program number
N010	G20	Inch programming

N020   G00   X1.000   Z.100   M03
        Rapid to position #1.
N030   G01   Z-.250   F2.0
N040   G02   X2.000   Y2.000   I2.000   F10.0
        Circular groove cut from point 1 to 2.
N050   X2.000   Y-2.000   J-2.000
        Circular groove cut from point 2 to 3.
N060   X-2.000   Y-2.000   I-2.000
        Circular groove cut from point 3 to 4.
N070   X-2.000   Y2.000   J2.000
        Circular groove cut from point 4 to 5.
N080   G00   Z.100
N100   M30   End of program
%

FULL CIRCLE PROGRAMMING

Most modern control systems are capable of programming a full circle (360°) that is a fairly common routine for operations such as:

- Circular pocket
- Spotface milling
- Helical milling (with linear axis)
- Milling a cylinder, sphere, or cone

Control systems that support the radius designation by the address **R** will also accept the **IJK**
modifiers but the reverse is not true. If both the
arc modifiers **IJK** and the radius **R** are programmed in the same block, the **R** (radius) value
takes priority and will be acted upon first. The
control that accepts only the modifiers **IJK** will
return an error message if the circular interpolation block contains the address **R** (an unknown
address). When using only the **R** address in programming 360°, it is wise to include either an **I** or
J address to ensure the best accuracy in full circle
programming.

Full circle cutting is defined as a circular tool
motion that completes 360° between the start and
end points which have identical coordinates.
Line N050 of Fig. 5-8-10 shows the **R** address and

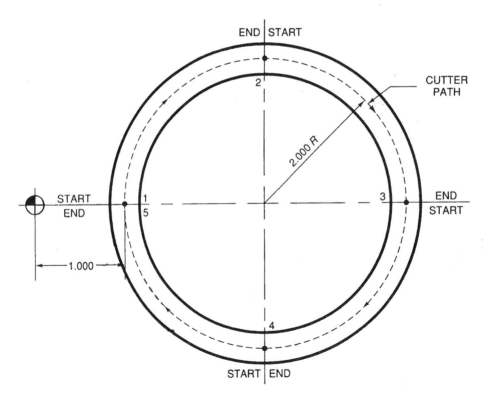

**Fig. 5-8-9 In center–point programming for a full circle, the start and end point of each quadrant (90°) must be
programmed.** *(Kelmar Associates)*

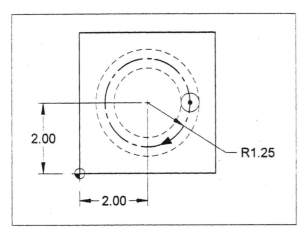

Fig. 5-8-10 Full-circle (360°) programming where the start and end point coordinates are the same. *(Peter Smid)*

the I and J coordinate modifiers in the program line.

```
O3456
N010  G90  G20
N020  G54  G00  X3.25  Y2.0  S800  M03
N030  G43  Z.100  H01  M08
N040  G01  Z-.250  F10.0
N050  G02  X3.25  Y2.0  I-1.25  J0  F12.0 (full
                        circle)
N060  G00  Z.100
```

SUMMARY

- Circular interpolation overcomes the difficulty of programming arcs and circles to make cutting of circular arcs and radii a routine operation.
- The main elements of a circle used in CNC programming are the radius, diameter, quadrants, and quadrant points.
- Center-point programming requires that each quadrant of a circle be programmed. With radius programming it is possible to machine a full circle in one program line.
- The information required for circular interpolation is the direction of cutter travel, the arc start and end points, and the center point of the arc or radius value.
- The radius of an arc can be designated with the R address or with the arc center modifiers I and J for the Mill, and I and Z for the lathe.
- Arc center modifiers I, J, and K are used to accurately describe the distance from the start point of an arc in relation to the center point of the arc.
- The information required for radius programming is the G-code, the end point of the arc, and the radius; for center-point programming, G-code, end point of arc, and center point of radius.
- Full Circle Programming requires the R address and the modifiers I, J, or K for the best results.

KNOWLEDGE REVIEW

1. For what purpose was circular interpolation developed?
2. How does the MCU break up an arc for circular interpolation?

Circle Elements

3. Name four main elements of a circle.
4. Define the diameter of a circle.

Programming Arcs

5. How do radius programming controls differ from center-point controls?

6. What three pieces of information are necessary for circular interpolation programming?
7. Define G02 and G03 codes.
8. Explain the relationship of the I and J modifiers to the X and Y axes.

Arc Center Information

9. G02 and G03 codes are modal; how are they replaced?
10. Why are arc center modifiers used?
11. Define the arc center modifier I.
12. Define the arc center modifier K.

Program Arc (Center-point)

13. What three questions should be answered before programming an arc?
14. List the information required for
    (a) center-point programming
    (b) radius programming
15. On machines that can only generate a 90° arc at a time, what information is required to program a full circle?

SECTION 6

Simple Programming

Unit 9 Programming Data

Unit 10 Reference Points, Compensation

SIMPLE PROGRAMMING

Simple programming consists of taking information from a part drawing and converting it into a form that can be transferred to a computer through the keyboard in a language designed for computer numerical control programming, Fig. 6-1. A program consists of a sequence of coded instructions of words, letters, symbols, etc., which cause a computer or CNC machine to carry out specific machining processes. The computer part program contains information about the machine and tools to be used, the details of the part required, and the path the cutter should take to machine the part. This computer program can then be sent directly to the machine tool for it to perform one or a series of operations or be stored for use any time in the future.

The program is generally prepared by a person who is familiar with computer numerical control (CNC) language. The person who can take a part drawing (or print) of a workpiece, determine what sequence of machining operations are required, and prepare a computer numerical control (CNC) program is called a *part programmer*. To properly prepare a program for any CNC machine, the programmer must be familiar with the language that is common to CNC and the machine control unit (MCU). The part programmer should also be familiar with the proper metal-cutting processes and sequences and the tools required for each operation, and be able to communicate this information accurately to the controls that operate the machine tool. A good programmer would be a machinist who developed these skills over years of training.

Fig. 6-1   The basic steps in programming a part *(Deckel Maho, Inc.)*

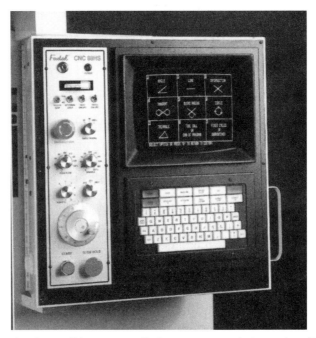

Fig. 6-2 The MCU makes it possible to manually input program information *(Fadal Engineering, Inc.)*

Programming at the Machine Tool

It is sometimes desirable or necessary to enter CNC data or revise the program at the machine tool. Modern CNC machine tools are equipped with controllers that have the capabilities to input data manually making program editing and correcting a simple operation, Fig. 6-2. It is thus possible to quickly edit programs or correct errors to respond to design changes and customers' requests. Modern MDI is an integrated programming system that incorporates a keyboard for programming, a video screen for displaying each program step, simulating the cutting tool motions, and reviewing the machining process to avoid errors that could result in damage to the CNC machine, cutting tools, or the workpiece. As microprocessors become more versatile and faster, with larger data storage, it will be possible to program more complex workpieces directly at the machine tool.

Unit 9
Programming Data

Manual offline programming of CNC machines is the creation of a part program that is coded using symbolic language that the MCU on the machine tool understands. The program is created by a part programmer who must list all the information and each specific operation required for the CNC machine to produce the part. The use of a computer or computer-aided programming equipment greatly reduces the programmer's work because it can receive and analyze data, check for input errors, make all necessary calculations, and lock in data in a logical sequence. The part programmer must refer to the part drawing and be able to plan the machining sequence and operations, the tools required for the job, draw up a suitable work plan, and present all this information in a language the MCU understands.

OBJECTIVES

After completing this unit you should be able to:

1. Understand the codes and functions required for simple CNC programming.
2. Prepare a CNC manuscript from the information on a part print.
3. Write a simple program for machining a part

KEY TERMS

axis presets	manuscript
continuous path	point-to-point positioning
cutter-diameter compensation	rapid traverse
fixed zero (home position)	tool length compensation

POINT-TO-POINT OR CONTINUOUS PATH

CNC programming falls into two distinct categories, point-to-point positioning and continuous path positioning, Fig. 6-9-1. The difference between the two categories was once very distinct. Now control units are able to handle both point-to-point and continuous path machining. A knowledge of both programming methods is necessary to understand what applications each has in CNC.

Point-to-Point Positioning

Point-to-point positioning is used when it is necessary to accurately locate the spindle, or the workpiece mounted on the machine table, at one or more specific locations to perform such operations as drilling, reaming, boring, tapping, and punching, Fig. 6-9-2. Point-to-point positioning is the process of positioning from one coordinate (**XY**) position or location to another, performing a machining operation, and continuing this pattern until all the operations have been completed at all programmed locations.

CNC drilling machines or any other CNC machine that must quickly move the workpiece from one position to another and perform an operation at that specific location are ideally suited for point-to-point positioning. For example, positioning a cutting tool such as a drill to an exact location or point, performing the drilling operation, and then moving to the next location (where another hole could be drilled). As long as

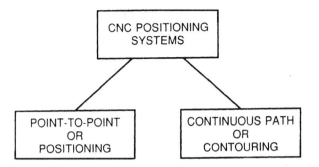

Fig. 6-9-1 Types of CNC positioning systems. *(Kelmar Associates)*

each point or hole location in the program is identified, this operation can be repeated as many times as required.

Point-to-point machining moves from one point to another at a rapid traverse rate, usually between 1000 to 2000 in./min.,(25.4 to 51 m/min) while the cutting tool is clear of the work surface. Rapid travel is used to quickly position the cutting

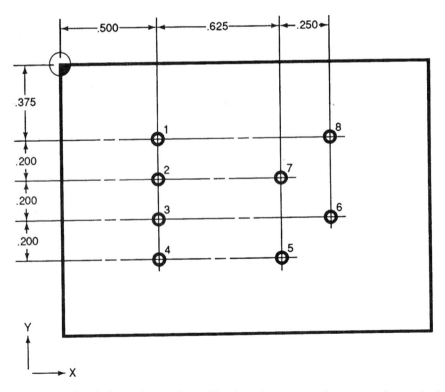

Fig. 6-9-2 The programmer takes information and specifications from a part drawing and records the machining sequence, tools required, etc., on a program sheet or manuscript. *(The Superior Electric Company)*

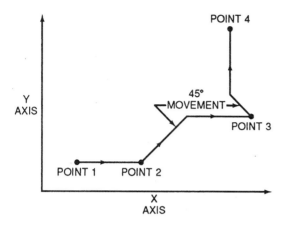

Fig. 6-9-3 The path followed by point–to–point positioning to reach various programmed points (machining locations) on the XY axis. *(Kelmar Associates)*

tool or workpiece between each location point before a cutting action is started. Movement on both **XY** axes is simultaneous and at the same rate during rapid traverse. The result is a movement along a 45° angle line until one axis is reached, and then there is a straight-line movement to the other axis. In Fig. 6-9-3, point 1 to point 2 is a straight line, and the movement is only along the **X** axis; but points 2 and 3 require that motion along both the **X** and the **Y** axes take place. As the distance in the **X** direction is greater than in the **Y** direction, **Y** will reach its position first, leaving **X** to travel in a straight line for the remaining distance. A similar motion takes place between points 3 and 4.

CONTINUOUS PATH (CONTOURING)

Contouring, or continuous path machining, can be used on work that is produced on a lathe or milling machine, where the cutting tool is in constant contact with the workpiece as it travels from one programmed point to the next. Continuous path positioning is the ability to control motions on two or more machine axes simultaneously to keep a constant cutter-workpiece relationship. The information in the CNC program must accurately position the cutting tool from one point to the next and follow a predefined accurate path at a programmed feed rate in order to produce the form or contour required, Fig. 6-9-4.

The method by which contouring machine tools move from one programmed point to the next is called **interpolation**. This ability to merge individual axis points into a predefined tool path is built into today's MCUs. There are six methods of interpolation: linear, circular, helical, parabolic, cubic, and NURBS. All contouring controls provide linear interpolation, and most controls are capable of both linear and circular interpolation. Helical, parabolic, cubic, and NURBS interpolation are used by manufacturers that produce parts with complex shapes, such as aerospace parts and car body dies.

Linear Interpolation

Linear interpolation consists of any programmed points linked together by straight lines, whether the points are close together or far apart, Fig. 6-9-5. Curves can be produced with linear interpolation by breaking them into short, straight-line segments. This method has limitations,

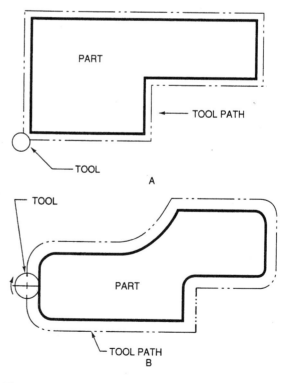

Fig. 6-9-4 Types of contour machining (A) Simple contour, (B) Complex contour. *(Allen Bradley)*

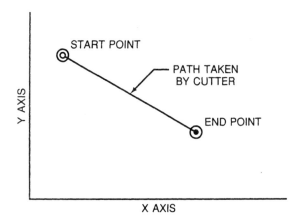

Fig. 6-9-5 An example of two–axis linear interpolation. *(Kelmar Associates)*

because a very large number of points would have to be programmed to describe the curve in order to produce a contour shape.

A contour programmed in linear interpolation requires the coordinate positions (**XY** positions in two-axis milling work) for the start and finish of each line or segment. Therefore, the end point of one line or segment becomes the start point for the next segment, and so on, throughout the entire program.

The accuracy of a circle or contour shape depends on the distance between each two programmed points. If the programmed points (**XY**) are very close together, an accurate form will be produced. To understand how a circle can be produced by linear interpolation, refer to Fig.

6-9-6. Figure 6-9-6B shows that when 8 connecting lines are machined, an octagon is produced, while Fig. 6-9-6C shows 16 connecting or chord lines. If the chords in each are examined, it should be apparent that the more program points there are, the more closely the form resembles a circle. If there were 120 chord lines, it would be difficult to see that the circle was made up of a series of short chord lines. Therefore, on a control unit that is capable only of linear interpolation, very accurate contours require very long programs because of the large number of points that need to be programmed.

Circular Interpolation

The development of MCUs capable of circular interpolation has greatly simplified the process of programming arcs and circles. To **program an arc, Fig. 6-9-7,** the MCU requires **the coordinate positions (XY axes) of the circle** center, the radius of the circle, the start point and end point of the arc being cut, and the direction in which the arc is to be cut (clockwise or counterclockwise), see Fig. 6-9-8. The information required may vary with different MCUs.

The circular interpolator in the MCU breaks up the distance of each chord line (circular span) into a series of the smallest movement increments produced by one single output pulse. This distance is usually .0001 in. or 0.001 mm, and the interpolator automatically computes enough

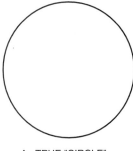

A - TRUE "CIRCLE"

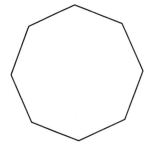

B - EIGHT SEGMENT "CIRCLE"

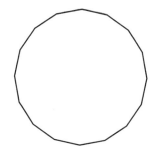

C - SIXTEEN SEGMENT "CIRCLE"

Fig. 6-9-6 Linear interpolation with arcs and circles. **(A)** True circle; **(B)** eight–segment circle; **(C)** sixteen–segment circle. *(Kelmar Associates)*

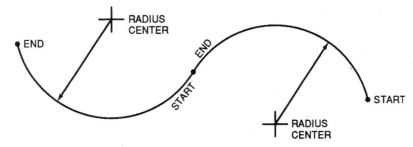

Fig. 6-9-7 Any complex form on two axes can be generated by circular interpolation. *(Kelmar Associates)*

output pulses to describe the circular form and then provides the control data required for the cutting tool to produce the form.

The advantage and power of circular interpolation is better appreciated by comparing it with linear interpolation. More than a thousand blocks of information would be needed to produce a circle in linear interpolation segment by segment. Circular interpolation requires only five blocks to produce the same circle, and some machines can do it in a single command.

Helical Interpolation

Helical interpolation combines two-axis circular interpolation with a simultaneous linear movement in the third axis. All three axes move at the same time to produce the helical (spiral) path required. Helical interpolation is most commonly used in milling large internal threads or helical forms, Fig. 6-9-9A.

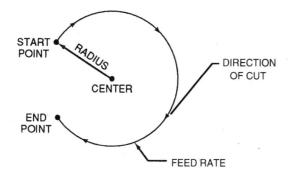

Fig. 6-9-8 For two–dimensional circular interpolation the MCU must be supplied with the XY axis, radius, start point, end point, and direction of cut. *(Kelmar Associates)*

Parabolic Interpolation

Parabolic interpolation has its greatest application in automotive dies, mold work, and any form of sculpturing. Parabolic interpolation is a method of creating a cutter path covering a wide variety of geometric shapes such as circles, ellipses, parabolas, and hyperbolas, Fig. 6-9-9B. Parabolic interpolation can be defined as a movement that is either totally parabolic or part parabolic and has three non-straight-line locations (two end points and one midpoint). Curved sections can thus be closely simulated, using about 50 times fewer program points than would be required using linear interpolation. The powerful computers of today have put parabolic interpolation in the background because it is now possible to use simpler interpolation.

Cubic Interpolation

Cubic interpolation allows complex cutter paths to be generated to cut the shapes of the forming dies used in the automobile industry. These dies can be machined with only a small number of input data points in the program. Besides describing the geometry, cubic interpolation smoothly blends one curved segment into the next without the interruption of boundary (start and stop) points. However, a larger-than-normal computer memory is required for cubic interpolation.

NURBS interpolation

The acronym "NURBS" stands for Non-Uniform Rational B-Spline. NURBS enables machines to cut complex shapes in smooth and continuous motions rather than approximating the shapes with a large number of short linear cuts.

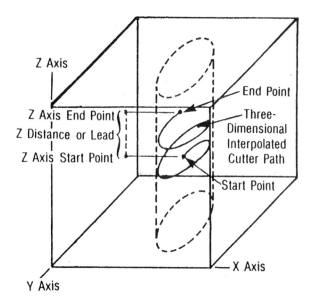

Fig. 6-9-9A Helical interpolation uses circular interpolation for two axes along with a linear movement in the third axis. *(Modern Machine Shop)*

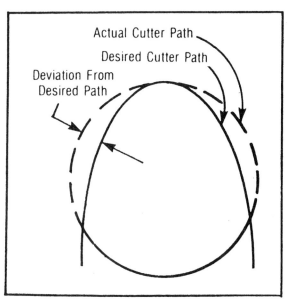

Fig. 6-9-9B Circular interpolation can be used to produce a curve on surfaces that are not true arcs. *(Modern Machine Shop)*

CAD systems have been using NURBS to define free-form curves and complex surfaces while controlling the smoothness of the form. NURBS has become the standard for surface representation in modern CAD/CAM systems. The primary advantages of NURBS is that it allows more complex surfaces to be defined and provides more precise data exchange between different CAD/CAM systems.

Interpolation Summary

The function of interpolation is to store programmed information, monitor and direct the machine axis motions, and keep a straight-line motion between coordinate points at a defined vector feed rate. The purpose of nonlinear interpolation is to eliminate the need to program numerous coordinate points for curved surfaces, which would be necessary with linear interpolation. Modern control units have the mathematical ability to generate the numerous points needed to produce curved surfaces if they are provided with a description of the curve.

MANUSCRIPT

Before a program for any workpiece for a CNC machine tool is started, the part print should be studied carefully. The surfaces of the workpiece that must be machined should first be noted, then the operations required, the sequence of machining operations, and the dimensional tolerances imposed. It is wise to remember that the machining of a part, by conventional machining or by CNC machining, is basically the same. In conventional machining, a skilled operator moves the machine slides manually, and in CNC machining the machine slides are moved automatically from the information supplied by the CNC program.

Manuscript Data Information

It is the job of the programmer to see that the machine tool receives the proper information to cut the part to the proper shape and size. Using numerical language, the programmer must record on a prepared form **(manuscript)** all the instructions that the machine tool must have in order to complete the job. The manuscript should contain all the machine tool movements, the cutting tools required, speeds, feeds, and any other information that might be required to machine the part. This information should be in a uniform format and be as clear as possible to give the CNC machine operator a good understanding of what is required. Figure 6-9-10 shows the types of information that should be included on or with a manuscript and should be supplied to the CNC operator by the programmer.

1. Part Sketch

A rough sketch should be made of the part and either incremental or absolute dimensions, or both if necessary, should be given for each axis location from the part zero (the most common), tool change point, or machine reference point.

2. Zero (or Reference) Point

- A zero or reference point should be established to permit the alignment of the workpiece and the machine tool, Fig. 6-9-11.
- A tool change position should also be selected that will leave enough room to change cutting tools and load and unload parts easily and quickly.

3. Workholding Device

- Always select the workholding device or fixture that can hold the part securely and not interfere with the machining operations.

- The fixture should not have any of its components too high above the part to be machined.
- The setup instructions for the fixture should be included in the manuscript.

4. Sequence of Operations

- Select the operations to be performed first and then the sequence of the following operations so that the part is machined to size and shape in the shortest time possible.
- Knowledge of basic machining operations is needed to list the operation sequence properly to produce an accurate part.

5. Axis Dimensions

- All the data necessary for every movement of the table or the cutting tool must be listed.
- This data must include axis locations for every surface to be machined and every hole to be drilled, tapped, reamed, etc.

6. Tool List and Identification

- Whenever tools are required, they should be indicated in the **Remarks or Comment** column of the manuscript.
- The tool identification number should indicate the sequence in which it will be used to machine the part, the tool diameter and length, and the tool and offset number.

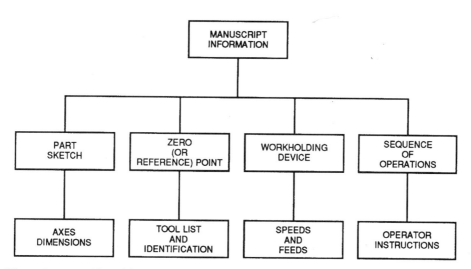

Fig. 6-9-10 The points considered in a manuscript to provide accurate information for the programmer and the CNC machine operator. *(Kelmar Associates)*

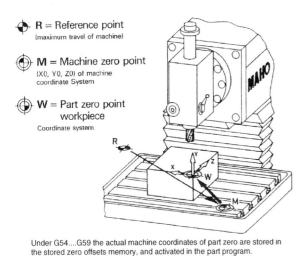

R = Reference point
(maximum travel of machine)

M = Machine zero point
(X0, Y0, Z0) of machine
coordinate System

W = Part zero point
workpiece
Coordinate system

Under G54....G59 the actual machine coordinates of part zero are stored in the stored zero offsets memory, and activated in the part program.

Fig. 6-9-11 The relationship between the part zero and the machine coordinate system. *(Deckel Maho, Inc.)*

PROGRAMMING PROCEDURE

The part illustrated in Fig. 6-9-12 will be used to introduce simple programming in easy-to-understand steps. Each step in the programming procedure will be explained in detail to provide the reader with a clear understanding of the meaning of the various codes and axis movements, and what happens as a result of each programming step.

Two programs (one in absolute and the other in the incremental mode) will be written, first to trace the part boundary and, second, to locate the positions of the holes. NOTE: Since this is only a programming exercise, the cutting tool will be .100 in. (**Z** axis) above the workpiece and the spindle will not be revolving.

Another program will be written in both the absolute and incremental mode to drill the six .375 in. (9.5 mm) diameter holes. In practice, programming for the hole locations and the drilling operation would be combined in the same program.

Notes

1. All programming begins at the zero or reference point (**XY** zero), which is located at the left of the part. This arrangement allows clearance for changing cutting tools and loading or unloading parts.

2. There will be no cutting on this first programming exercise, so the tool will be set at Z.100 above the part surface.
3. The part boundary will be programmed clockwise, starting at point A.
4. Return to **XY** zero.
5. Program the hole locations.
6. Return to **XY** zero.

CODES NEW TO THIS UNIT

G80	Cancels canned cycle
G81	Fixed or drilling canned cycle
M00	Optional stop (manual tool-change machines)
M06	Spindle OFF
R	Gage height used in cycles

The Part Boundary Program (Absolute)

O9870			Program number
N010	G90	G20	
	N010		Sequence numbers should be in progressions of 10 to leave room for other steps to be inserted if necessary.
	G90		Absolute program mode
	G20		Inch input mode
N020	G00	G54	X1.000
	G00		rapid traverse rate.
	G54		work offset
	X1.000		cutting tool located 1.000 to the right along **X** axis to point A.
N030	G43	Z.100	H01
	G43		Tool length offset
	Z.100		Tool .100 above work surface
	H01		Tool length offset
N040	G01	Y3.625	
	G01		linear interpolation (straight-line movement)
	Y3.625		moves 3.625 up along the **Y** axis to point B.
N050	X7.000		

	X7.000	Moves 6.000 to the right along the **X** axis to point C (7.000 from the **XY** zero).
N060	Y0	
	Y0	Moves down 3.625 along the **Y** axis to point D.
N070	X1.000	
	X1.000	Moves 6.000 to the left along the **X** axis back to point A.
N080	G00 X0	Returns to **XY** zero in the rapid mode.

Hole Locations Program (Absolute)

N090 X2.000 Y.875

Rapids 2.000 along the **X** axis and .875 up the **Y** axis to hole location #1 because it is still in the rapid mode from sequence number N080.

N100 X4.000

Rapids 2.000 along **X** axis to hole #2 (4.000 to the right of the **XY** zero).

N110 X6.000

Rapids 2.000 (6.000 from **XY** zero) along **X** axis to hole #3.

N120 Y2.750

Rapids 1.875 (2.750 from **XY** zero) up the **Y** axis to hole #4.

N130 X4.000

Rapids 2.000 to the left to hole #5.

N140 X2.000

Rapids 2.000 to the left to hole #6

N150 X0 Y0

Rapids back to **XY** zero.

N160 M30

End of program.

%   Rewind code.

Part Boundary Program (incremental)

O9871

N010 G90 G20

| G90 | Absolute program mode. (always used in incremental programming to start the program and move to the first location) |

N030 G00 G54 X1.000 Z.100

G00	rapid traverse mode.
X1.000	cutting tool located 1.000 to the right along **X** axis to point A.
Z.100	Tool .100 above work surface

N040 G01 G91 Y3.625

| G01 | linear interpolation (straight-line movement). |

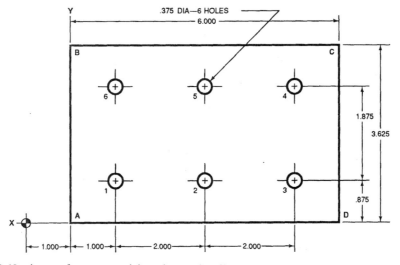

Fig. 6-9-12   A sample part requiring six .375 in. diameter holes to be drilled. *(Kelmar Associates)*

Y3.625   moves 3.625 up along the
                Y axis to point B.
G91   Incremental program mode
N050 X6.000
      Moves 6.000 to the right along X axis to
      point C.
N060 Y-3.625
      Moves 3.625 down along the Y axis to
      point D.
N070 X-6. 000
      Moves 6.000 to the left along X axis
      back to point A.
N080 G00 X-1.000
      Returns to XY zero in the rapid mode.

Hole Locations Program (incremental)

N090 X2.000   Y.875
      Rapids 2.000 along X axis and .875 up Y
      axis to hole location #1 because it is still
      in the rapid mode from sequence
      number N080.
N100 X2.000
      Rapids 2.000 along X axis to hole #2.
N110 X2.000
      Rapids to hole #3.
N120 Y1.875
      Rapids 1.875 up Y axis to hole #4.
N130 X-2.000
      Rapids to left 2.000 along X axis to hole #5.
N140 X-2.000
      Rapids to left 2.000 along X axis to hole #6.
N150 X-2.000   Y-2.750
      Rapids back to XY zero.
N160 M30
%   Rewind code.

Program #2-Drilling Holes

A review of the R work plane, the Z axis motion,
and fixed or canned cycles covered in Unit 4-6
helps make clear how the holes for the workpiece
shown in Fig. 6-9-12 can be drilled as efficiently
as possible.

• The **R work plane**, or clearance height, is generally .100 in. above the surface of a workpiece
and is used as a reference, and all other work
surfaces are relative to this location.
• The **Z axis** motion moves the cutting tool either
into the workpiece (a minus motion) or away
from the workpiece (a plus motion).
• **Fixed or canned cycles** (G81 to G89) are preset
combinations of operations, such as drilling,
where all machine axis motions are programmed and will repeat themselves until canceled by a G80 code.

As can be seen from the following example,
all these factors can be programmed and will
repeat themselves until they are canceled by the
G80 code.

N040 G81 X2.000 Y1.500 R.100 Z-1.000 F5.0
      G81   A fixed or canned drilling
                   cycle.
      R.100   Gage height set at .100
                   above the work surface.
      Z-1.000   The drill is fed 1.000 deep
                   into the work.
      F5.0   The drill feed rate is set at
                   5.0 in./min.
      • After reaching the Z depth, the drill
        retracts automatically, in the rapid mode,
        out of the hole to the gage height (R level).

Drilling Program #2 (Absolute)

To drill the six .375 in. diameter holes illustrated
in Fig. 6-9-12, the following points should be kept
in mind:

Notes
    1. The machining operation is drilling.
    2. The zero or reference point (**XY** zero) is
       located to the left of the part.
    3. The G81 fixed or canned drilling cycle will
       be used to eliminate repetition in pro-
       gramming.
    4. Both absolute and incremental program-
       ming will be explained.

O5432   Program number
N010 G90 G20
      G90   absolute mode

	G20	inch programming
N020	G00 G54 X-1.000 Y0 S1000 M03	
	G00	Rapid traverse mode
	G54	The work offset at −1.000 at left of part
	S1000	Spindle speed 1000 r/min.
	M03	Spindle **ON** clockwise
N030	G81 X2.000 Y.875 R.100 Z-1.000 F5.0 M03	
	G81	Drilling cycle
	X2.000)	
	Y.875)	Machine rapids to hole #1 location
	R.100	Spindle rapids to .100 above the work surface
	Z-1.000	The drill fed 1.000 in. into the part
	F5.0	5 in. drilling feed rate and then spindle rapids to .100 above the part
	M03	Spindle **ON** clockwise
N040	X4.000	

- The table rapids to hole #2 position 4.000 in. from the **XY** zero or reference point
- The G81 fixed cycle (N030) will be repeated, and a hole #3 will be drilled

N050   X6.000

- The table rapids to hole #3 position
- The G81 cycle is repeated

N060   Y2.750

- The table rapids to hole #4 position
- The G81 cycle is repeated

N070   X4.000

- The table rapids to hole #5 position
- The G81 cycle is repeated

N080   X2.000

- The table rapids to hole #6 position
- The G81 cycle is repeated

N090   G80

	G80	Cancels the drill cycle and automatically puts the machine in the rapid mode

N100   X0 Y0 Z.100 M06

- The table rapids back to the **XY** zero or reference point
- Z.100 raises the cutting tool to retract position.
- M06 stops the machine spindle

N110   M30   End of program

N120   M00

- Optional stop, used only on a manual tool change machine. Safety factor to prevent machine from starting while tools are being changed
- The cycle must be reactivated by pressing the cycle start control

%	Rewinds the program in preparation for use in drilling the next part

Drilling Program #2 (incremental)

The same part, Fig. 6-9-12, with six holes to be drilled, will be used to show the differences between absolute and incremental programming.

O5433

N010	G90 G20	
	Absolute programming, inch mode	
N020	G00 G54 X-1.000 Y0 M03	
	G00	Rapid traverse mode
	G54	The work offset at −1.000 at left of part
	M03	Spindle **ON** clockwise
N030	G81 X2.000 Y.875 R.100 Z-1.000 F5.0 M03	
	G81	Drilling cycle
	X2.000)	Machine rapids to hole #1
	Y.875)	location
	R.100	Spindle rapids to .100 above the work surface
	Z-1.000	The drill feeds 1.000 in. into the part
	F5.0	5 in. drilling feed rate set, then spindle rapids to .100 above the part
	M03	Spindle **ON** clockwise

- The drill rapids out of the hole back to gage height (.100 above work)

N040 G91 X2.000

- Incremental mode
- The table rapids 2.000 in. along the **X** axis to hole #2 position
- The G81 fixed cycle (N030) is repeated, and hole #2 is drilled at this position

N050 X2.000

- The table rapids 2.000 in. to hole #3 position
- The G81 cycle is repeated, and hole #3 is drilled

N060 Y1.875

- The table rapids 1.875 up along the **Y** axis to hole #4 position
- The G81 drilling cycle is repeated

N070 X-2.000

- The table rapids –2.000 along the **X** axis to hole #5 position
- The G81 cycle repeats to drill hole #5

N080 X-2.000

- The table rapids –2.000 along the **X** axis to hole #6 position
- The G81 cycle repeats to drill hole #6

N090 G80

- Cancels the drill cycle, raises the spindle above the work and automatically puts the machine in the rapid mode.

N100 G28 X-2.000 Y-2.750 M06

- G28 zero return in incremental mode
- The table rapids simultaneously along the **XY** axes and returns to the **XY** zero

M06 — Turns the machine spindle **OFF** and raises the cutting tool to the full retract position

N110 M30 End of program

% Rewinds the program in preparation for drilling the next part.

M00

- Optional stop, used only on a manual tool change machine.

- Safety factor to prevent machine spindle from starting while tools are being changed.
- The cycle must be reactivated by pressing the cycle start control.

CUTTER-RADIUS COMPENSATION

Since the basic reference point of the machine tool is never at the cutting edge of a milling cutter, the programmer must take the diameter of a cutter into consideration when programming. The center of the milling cutter is not the part that does the cutting; rather, it is some point on the periphery of the cutter. If a 1.000 in. end mill is used to machine the edges of a workpiece, the programmer would have to keep a .500 in. offset from the normal surface in order to cut the edges accurately, Fig. 6-9-13. The .500 in. offset represents the distance from the centerline of the cutter or machine spindle to the edge of the cutter that will be machining the edges. Therefore, whenever a part is being machined with some type of milling cutter, the programmer must always calculate an offset path, which is usually half the diameter of the cutter used.

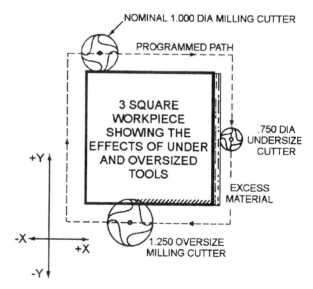

Fig. 6-9-13 **Compensation must be made for various milling cutter diameters when programming in order to cut to the proper dimensions.** *(Kelmar Associates)*

On modern MCUs that have part contour programming, the cutter centerline offsets are calculated automatically once the radius of the cutter for each operation is entered into the program. Many MCUs have operator-entry capabilities that can compensate for differences in cutter diameters; therefore an oversize cutter or one that has been sharpened can be used as long as the compensation for oversize or undersize cutters is entered.

AXIS PRESETS

Some MCUs have two position registers, the machine position register and the command register, Fig. 6-9-14. The **machine position register** displays a continual record of the position of the tool from the absolute zero, or home, position. This **XY** zero (or home) position of a machine is generally in the top right-hand corner of the table, though it can be in any corner depending on the manufacturer. The **command register** displays a continual record of the axis machine po-

sition in relation to the programmer's (or operator's) defined start position.

The programmer often uses an **XY** zero or reference point that is at some position away from the workpiece to allow for tool changes and loading or unloading workpieces. The home position of the machine table and the **XY** zero of the programmer often are not the same. Many MCUs have the capability of mathematically shifting the machine zero to any point within the working range of the machine table so that it matches the programmer's **XY** zero. The **zero shift control** on the MCU allows the **XY** zero position to be shifted from the machine's absolute zero (home position) to the programmer's **XY** zero.

An MCU equipped with **full floating zero** allows the operator to fasten a workpiece at any position on the machine table that is convenient. Once the workpiece is securely fastened, the alignment positions are entered during setup and only the program line number is entered into the machine controls. With the use of zero shift control or a dial, the **XY** zero of the manuscript is accurately located. The full floating zero pro-

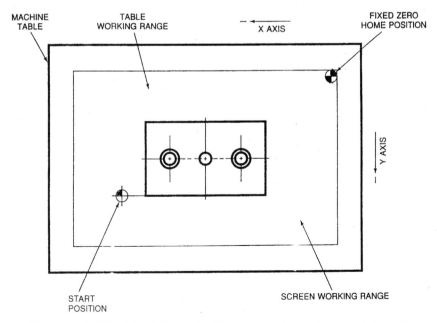

Fig. 6-9-14 The zero shift control of the MCU allows the XY zero location to be moved from the machine (home) position to any point within the working range of the table. *(Kelmar Associates)*

vides the operator with flexibility and greatly reduces the setup time because the **XY** zero can be located at any place on the machine table.

TOOL LENGTH COMPENSATION

A part programmer's initial planning must include the operation-by-operation sequence of how a part will be machined. At the same time, a list of cutting tools used for each operation must be developed. This *tool list* must include the type of each cutting tool and its diameter and length. Since a wide variety of cutting tools are generally used for machining a part, some allowance must be made for the differences in their length to ensure that machining operations are performed to the correct depths.

The programmer must specify a minimum length based on the maximum depth to which each cutting tool must travel. Each tool is then programmed as though it had **zero length,** and the spindle/tool gage line and the tool point were the same. The amount of Z axis travel is programmed .100 above the work surface at a specified feed rate. The length of each tool is entered into the MCU by either the CNC part programmer or the machine operator.

Some industries find it more convenient to preset all cutting tools to allow for differences in tool length, which results in consistency of length every time a particular tool is used. Other companies use **tool assembly drawings,** which describe the cutting tool and give the setting length for each tool. Each cutting tool is assigned a specific number, which is stored in the MCU and can be recalled any time that specific tool is used.

Some MCUs are equipped with **semiautomatic tool compensation** to make allowances for differences in cutting tool lengths. The tool length compensation feature provides for as many as 64 or more tool length offsets. The programmer uses the same basic tool length for all tools. When the machine is being set up, the operator mounts each tool, generally starting with the longest, sets it against a fixed reference point, and presses a button on the MCU. This automatically records the setting in the memory. All the cutting tools for a particular job are set against the same reference point, and the difference between each tool's actual length and the basic tool length is entered into memory as a compensation value, Fig. 6-9-15. Individual offsets are activated or recalled from memory by H words.

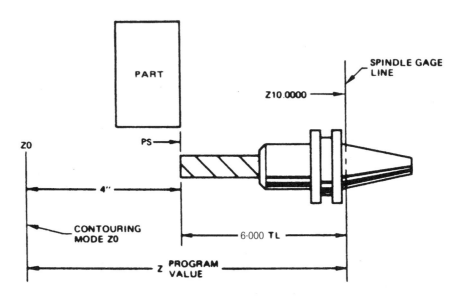

Fig. 6-9-15   The tool "rapids" to the gage height and then travels at a set feed rate to the Z depth.

SUMMARY

- The use of a computer for CNC programming reduces a programmer's work because it can receive and lock in data in a logical sequence.
- Point-to-point positioning is the process of positioning a machine spindle quickly from one position to another while the cutting tool is clear of the work.
- Continuous path or contouring machining can control the motions of two or more machine axes to follow a programmed accurate path at a programmed feed rate.
- Linear interpolation consists of any straight-line motion along the X, Y, and Z axes to perform a machining operation.
- Circular interpolation is used to program and machine arcs and circles, clockwise or counterclockwise, with all the necessary calculations being made by the interpolator in the machine control unit (MCU).
- A CNC program manuscript should contain all the information about a part so that the CNC machine can produce the shape and accuracy required.
- Programming must include programming modes, inch/metric input, zero or reference point locations, and part specifications.
- Fixed or canned machining cycles are used for repetitive machining operations, such as drilling because any operation can be repeated by a code in the program.
- Centerline offsets are calculated automatically if the diameter of a cutter used is entered into the program.
- Tool length compensation is used to make allowances for the different lengths of cutting tools used for machining a part.

KNOWLEDGE REVIEW

1. List five advantages of computer or computer-aided programming.

Point-to-Point or Continuous Path

2. Name the two CNC positioning systems and state the purpose for which they are used.
3. What is the purpose of rapid traverse, and when is it used?
4. At what angle does rapid traverse travel when movement along the XY axis is required simultaneously?

Continuous Path (Contouring)

5. What is *continuous path positioning (contouring)?*
6. Define *interpolation.*
7. For what purpose is linear interpolation used?

Interpolation

8. Why is linear interpolation not often used to produce accurate arcs and contours?

9. What four pieces of information are required to program an arc by circular interpolation?
10. How small is the distance of each chord line of the circular interpolator in the MCU?
11. What axis movement occurs in helical interpolation?
12. For what purpose is parabolic interpolation used?
13. Why is cubic interpolation widely used in the automotive industry?

Manuscript

14. Explain the difference between machining a part by conventional machining and by CNC.
15. What information should be included on a manuscript?
16. What is the purpose of the zero or reference point?
17. Why should the information regarding the tool list and identification be included on a manuscript?

Programming Procedure

18. Why is the XY zero or reference point usually located off the edge of a part?
19. Explain the difference between the absolute and incremental programming modes.
20. Define the following codes:
    (a) G91, (b) G20, (c) G00, (d) M30
21. Use a diagram to illustrate the direction of movement of the following:
    X+, X–
    Y+, Y–

Drilling Holes

22. Define the R work plane and the Z axis motion.
23. What are fixed or canned cycles and why are they useful in CNC programming?

24. Describe what happens in a complete fixed or canned cycle.

Cutter Diameter Compensation

25. Why is cutter diameter compensation important to programming operations involving milling cutters?
26. How do modern MCUs compensate for oversize and undersize cutter diameters?

Tool Length Compensation

27. What information should be included on a tool list?
28. How can the actual tool length be entered into the MCU?
29. Why do some industries use preset tooling?
30. What is the purpose of tool assembly drawings?

Unit 10

Reference Points, Compensation

CNC programmers work in a very precise mathematical condition where the machine tool, the part or workpiece, and the cutting tool must work together to produce a part. These three items must have some common factor and be able to work together to produce an accurate part in the correct location. This common factor is a reference or starting point that is a fixed or selected location, along two or more axes, on a CNC machine, the cutting tool, or the part. It is the coordinate location from which all measurements are taken to machine a part. Without a reference point, it is possible that a part could be machined:

- To the exact size and shape required, but in the wrong location.
- In the right location but to the wrong size or shape.

All three components must work in harmony with each other to produce a part in the correct location and to the size required.

OBJECTIVES

After completing this unit, you should be able to:

1. Select a tool-change position that is convenient for changing tools quickly without interfering with the workpiece.
2. Locate the workpiece zero accurately at the edges or specific location on the part.
3. Move the workpiece zero to any point convenient for the machining operations required.
4. Supply the cutting tool data necessary to machine the part to an accurate size.

KEY TERMS

datum	tool-change point
machine zero (origin)	tool-length compensation
part zero	tool-radius compensation
reference point	work coordiante system
stored zero shift	zero point offset

MACHINE AND WORK COORDINATES

Most CNC machine tools have a default coordinate system (zero or home position). This is a fixed position, usually located at the tool change position, which is built into the machine by the manufacturer. It is used as a reference point when locating work-holding devices and workpieces. Most MCUs have two position registers, the machine position and the command position, Fig. 6-10-1. The absolute position register shows the position of the machine table from the machine zero or home position at all times. The offset register shows the table axis position in relation to the programmer's or operator's XY zero or reference point.

When a part is being programmed, the machine's zero position and the part zero are rarely the same. The programmer can choose any location on the part, workholding device, or machine table as the program zero.

REFERENCE POINTS

A reference point is a fixed or selected location along two or more axes on a machine, part, or cutting tool that is used as a starting point for setup or machining operations. Figure 6-10-2 shows the three main reference points used on a CNC mill for programming:

Technical Terms	Commonly Used Terms
1. **Machine** reference point	machine zero or home
2. **Workpiece** reference point	program zero or part zero
3. **Tool** reference point	tool tip or command point

MACHINE REFERENCE POINT

The **machine zero point**, commonly referred to as the **machine zero, home zero,** or **machine reference point**, is the zero point of the machine's coordinate system. Its location varies with manufacturers but is usually at the positive end of the travel range of each axis. The machine zero is built into the CNC machine at the time of manufacture and cannot be changed by the operator. Figure 6-10-3A and 3B show the machine zero

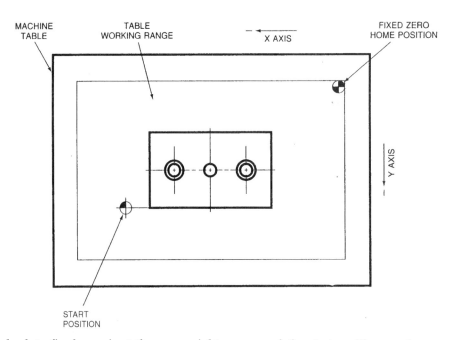

Fig. 6-10-1 The absolute fixed zero is at the upper right corner and the start position can be moved to any convenient location. *(Kelmar Associates)*

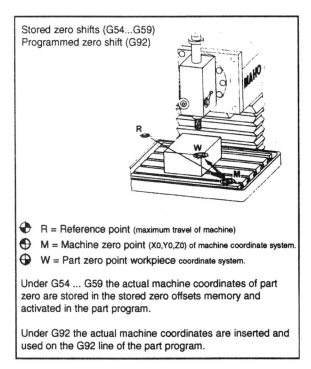

R = Reference point (maximum travel of machine)

M = Machine zero point (X0,Y0,Z0) of machine coordinate system.

W = Part zero point workpiece coordinate system.

Under G54 ... G59 the actual machine coordinates of part zero are stored in the stored zero offsets memory and activated in the part program.

Under G92 the actual machine coordinates are inserted and used on the G92 line of the part program.

Fig. 6-10-2 The relationship between the home or machine zero position and the part zero. *(Deckel Maho, Inc.)*

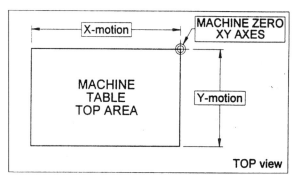

Fig. 6-10-3A A top view of a vertical machine zero XY axes at the upper right corner of the table. *(Peter Smid)*

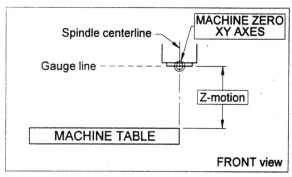

Fig. 6-10-3B A front view of a vertical machine Z axis in the home position. *(Peter Smid)*

point for a CNC vertical-machining center from two views.

> Machine zero is a fixed position on a CNC machine that can be reached at any time through the control system (MCU), manual data input (MDI), or the program execution code.

manual for the machine zero when operating a new or different machine.

Turning Centers

On turning centers the machine reference point(**R**) is always at the furthest limit of travel **away from the spindle centerline**, Fig. 6-10-4.

Machining Centers

The most common and standard machine reference point for vertical machining centers is usually in the **upper right corner** of the machine, Fig. 6-10-3A. Since manufacturers can vary this location, machine zero may be found at the:

- Lower left corner of the machine——(X–Y–)
- Upper left corner of the machine——(X–Y+)
- Lower right corner of the machine——(X+Y–)
- Upper right corner of the machine——(X+Y+)

It is always wise to check the CNC machine

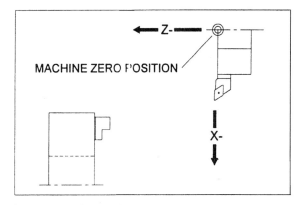

Fig. 6-10-4 The machine zero position for a typical CNC lathe with rear tooling. *(Peter Smid)*

The zero point for the coordinate system of the machine is always at the center of the spindle nose face (center line), Fig, 6-10-5A and 5B. The main spindle axis (centerline) represents the **Z** axis, the spindle face is the **X** axis. When a tool travels in a positive direction, it moves away from the workpiece.

WORKPIECE REFERENCE POINT

After the part has been aligned and fastened in a suitable workholding device, the reference point previously selected is used for programming purposes. This point is generally called the program zero or part zero and can be located anywhere that best suits the programmer and the

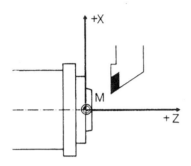

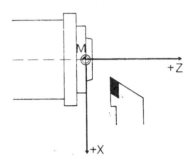

Fig. 6-10-5A The position of the machine zero (M) on CNC lathes when the cutting tool is behind the center line. *(Hanser Publishers)*

Fig. 6-10-5B The position of the machine zero (M) on CNC lathes when the cutting tool is in front of the center line. *(Hanser Publishers)*

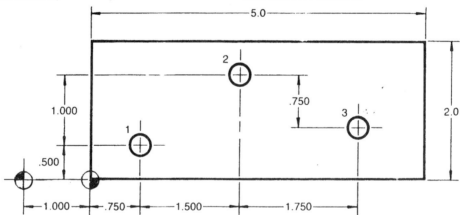

Fig. 6-10-6 The program or part zero in the lower left corner of the part. *(Kelmar Associates)*

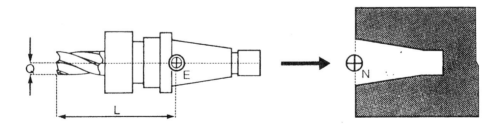

Fig. 6-10-7 The tool socket point (N) and the tool setting point (E) match when inserted in the tool adapter. *(Hanser Publishers)*

machining operations. Some points to consider when selecting the workpiece zero are:

- Is it the most convenient place at which drawing dimensions can be converted into coordinate values?
- Will this location produce the best accuracy and machining conditions for the part?
- Is this location suitable for easy setup operations?
- Will this location be safe for the setup and machine operator?

Machining Centers

The most common types of methods of holding workpieces on machining centers are vises, chucks, subplates, fixtures, tombstones and clamping directly to the machine table. The program or part zero on a workpiece can be set at any location that is convenient for programming or machining purposes. It is usually found in the lower left corner of the part, Fig. 6-10-6, however it can be at any point on the part, or the workholding device that is most suitable for the job or the machining operation.

When using preset tools, the tool-setting point (E) is located at a certain point on the toolholder and the corresponding point (N) is in the tool carrier or machine spindle, Fig. 6-10-7.

Turning Centers

The workpiece or program zero point for parts machined on CNC turning centers should be located on the spindle axis or centerline. Either the left-hand or right-hand end of the part can be

used for programming purposes but the right end is preferred, Fig. 6-10-8.

On CNC turning centers with multiple tool turrets or carriers, it may be necessary to use additional reference points such as the tool carrier reference point (T) and the slide reference point (F), Fig. 6-10-9. Preset tools used in the tool carrier would use point (E) on the toolholder and point (N) on the tool socket as reference points, Fig. 6-10-10.

TOOL SETTINGS and OFFSETS

All types of CNC machine tools require some form of tool setting and offsets (compensation) to allow the programmer or operator to meet unexpected problems relating to tooling. Usually it is impossible to predict the exact tooling problems that can arise on a job. Tool compensation allows changes to be made to the program to overcome the tooling problems that may arise.

Setting Zero Point and Tool Axis

Before any part is machined, it is very important to set the cutting tool to the zero locations of the **X, Y**, and **Z** axes. Each machine tool has a **home** or **zero reference point** that is built into it by the manufacturer and cannot be changed. In Fig. 6-10-1, the **XY** zero (home) position of the machine is in the upper right-hand corner of the working range of the table. However, the programmer can set any reference point that best suits the part, operation, or tool change position.

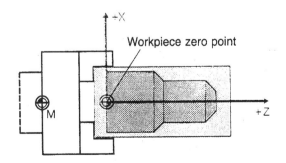 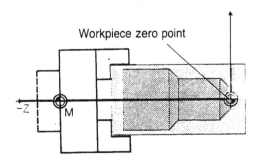

Fig. 6-10-8 The workpiece or part zero may be set at either the right–hand or left–hand edge of a part. *(Hanser Publishers)*

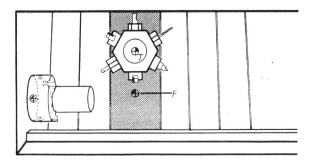

Fig. 6-10-9 The tool carrier and slide reference points of a CNC turning center. *(Hanser Publishers)*

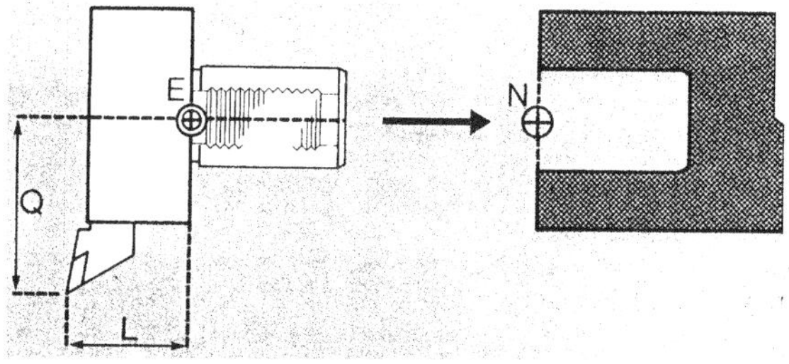

Fig. 6-10-10 The tool socket and toolholder reference points on CNC turning centers. *(Hanser Publishers)*

This reference point will usually not be the same as the machine reference point.

Setting X and Y Axes to Zero

After a flat part has been aligned and fastened to the table or some workholding device, the center of the machine spindle should be aligned with the **X** and **Y** edges. This can be done by:

- Setting an edge finder to the part's left edge, Fig. 6-10-11.
- Moving the table one-half the diameter of the edge finder along the X axis.

> The edge finder is .200 in. in diameter, therefore a movement of .100 in. in the direction of the arrow will align the spindle center with the workpiece edge.

- Set the **X** axis register to zero.

The same procedure should be used to set the **Y** axis, Fig. 6-10-12, and then the **Y** axis register is set to zero. The **zero shift** on the MCU (machine control unit) allows the **XY** zero location to be shifted to another one chosen by the programmer.

Setting the Z Axis to Zero

To set the end of the milling cutter to the surface of the workpiece in the **Z** axis, Fig. 6-10-13:

- Bring the cutter over the surface of the part.
- Place a thin piece of paper between the surface of the work and the end of the cutter.
- Lower the spindle or raise the table until a slight drag is felt on the paper.
- Move the machine table or spindle sideways until the cutter clears the workpiece edge.
- Raise the table or lower the spindle .002 in. to allow for the thickness of the paper.

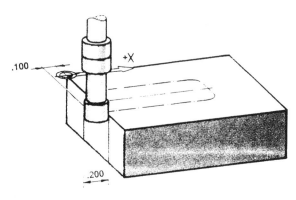

Fig. 6-10-11 Setting the spindle center to zero on the left edge of a part (X axis) with an edge finder. *(Deckel Maho, Inc.)*

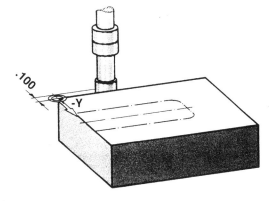

Fig. 6-10-12 Setting the spindle center to zero on the top edge of a part (Y axis) with an edge finder. *(Deckel Maho, Inc.)*

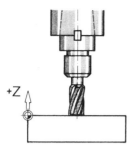

Fig. 6-10-13   Setting the end (bottom) of an end mill to the top of a work surface in the Z axis. *(Deckel Maho, Inc.)*

WORK SETTINGS and OFFSETS

All CNC machine tools require some form of tool setting, work setting, and offsets (compensation). Compensation allows the operator or programmer to make adjustments for unpredictable tooling and setup conditions. Sometimes it is almost impossible to predict certain conditions and compensation is especially designed to handle such problems.

Work Coordinates

In absolute positioning, work coordinates are generally set on one edge or corner of a part and all programming is taken from this work coordinate. In Fig. 6-10-14, the part zero is used for all positioning for hole locations #1, 2, and 3.

In incremental positioning, the work coordinates change because each location is the zero point for the move to the next location, Fig. 6-10-15.

On some parts, it may be desirable to change from absolute to incremental, or vice versa, at a certain point in the job. This can easily be done by inserting the G90 (absolute) or the G91 (incremental) command into the program at the point where the change is to be made.

Work or R plane

The word-address letter **R** refers to either the work surface or the rapid-traverse level (often called the **work plane**) programmed. The R level surface is set a specific height or distance above the work surface and it can be used with canned or fixed cycles. This setting is referred to as the **R level**, or the reference dimension, and all programmed depths for cutting tools and surfaces to be machined are taken from the **Z0** surface.

The **Z0** level plane is generally set at the highest surface of the workpiece, Fig. 6-10-16, which is also known as gage height.

When setting up for differences in cutting tool length, the operator generally places a shim or feeler gage on top of the highest surface of the workpiece. Each tool is lowered until it just

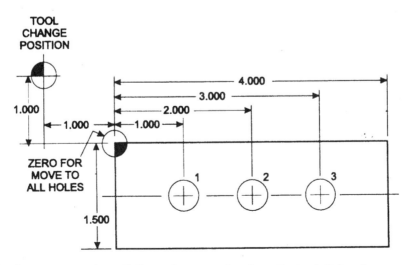

Fig. 6-10-14   In absolute programming, all dimensions are taken from the top left–hand corner of the part. *(Kelmar Associates)*

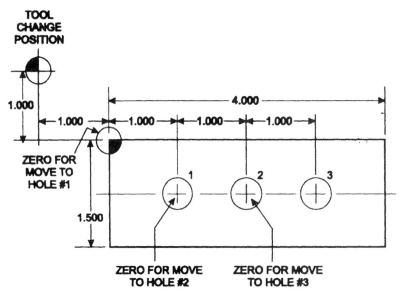

Fig. 6-10-15 In incremental programming, all dimensions are taken from the end of the previous point. *(Kelmar Associates)*

touches the gage surface and then its length is recorded on the tool list.

TOOL OFFSETS

For every part, the programmer must consider the step-by-step operations required to machine the part. At the same time, a list of cutting tools used for each operation must be included. This **tool list** must include the type of each cutting tool, its diameter, and length. Since a wide variety of cutting tools, of various lengths and diameters, are generally used for machining a part, it is important to be able to compensate for the differences in their diameters and lengths to en-

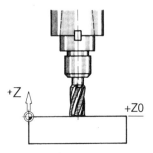

Fig. 6-10-16 The end of the milling cutter is set to the **Z0 on the top of the work surface.** *(Deckel Maho, Inc.)*

sure that the part will be machined accurately. This compensation involves working with offsets so that the machine control unit knows exactly how to adjust for differences in tool diameters and lengths.

All forms of compensation work with offsets. CNC offsets can be thought of as memories on an electronic calculator. If the calculator has memories, a constant value can be stored into each memory and used whenever required for a calculation. The need to enter the same number over and over again when it is required is thus avoided. Like the memories of an electronic calculator, offsets in the CNC control are storage locations into which numerical values for each tool can be placed.

Tool Lengths

A wide variety of cutting tools are used on CNC machining centers and the length of each tool consists of the cutting tool and the holder or adaptor in which it is mounted. Since milling tools vary in length, the CNC program must contain a tool-length compensation factor value for each cutting tool used in order to produce accurate parts. Differences in tool lengths may be set either at the machine (**on-machine**) or away from the machine (**off-machine**). To make these settings accurate and consistent, the gage line of the **Z** axis is used.

Gage Line

Most CNC machine tool manufacturers establish a reference position or gage line which is an imaginary line used for gaging or measuring along the **Z** axis, Fig. 6-10-17. This position has a fixed distance relationship with the top of the machine table and this distance cannot be changed. The relationship between the gage line and the surface of the table allows the operator to program the tool motion accurately along the **Z** axis. All tool length compensation uses the gage line as the basis for adjustment for variations in tool lengths.

TOOL LENGTH COMPENSATION

The four common methods of setting tool-length compensation are by **presetting tools, touch-off method, master tool method,** and **semi-automatic compensation**.

1. Presetting Tool Lengths

Most modern CNC machines use preset tools that have been preset in the toolroom because it reduces the amount of machine nonproductive time that is spent during setup. Their type, diameter, and length have all been recorded on the tool assembly drawing and are used when the program is being prepared. Each tool is assigned a specific identification number (**H**) along with a list of measured (preset) tool lengths. The CNC operator must set the tools in their correct location in the tool magazine and register each tool length in the offset register using the proper offset (H) number, Fig. 6-10-18. Any tool can be recalled by a command in the CNC program whenever a specific tool is required.

Touch Off Method

Touch off is a common method of setting tool lengths even though it does take away some of the machine's productive time. Each tool is assigned an H number, also called the tool length offset number, which usually corresponds to the tool number. The setup procedure is to measure the distance the tool travels from the machine zero position (home) to the program zero position (Z0), Fig. 6-10-19.

- This distance, always negative, is entered into the corresponding **H** offset numbers under the tool-length offset menu of the control system.
- The **Z** value for any work offset (G54-G59 and the common offset) is normally set to **Z0.**

Master Tool Method

The master tool method, using the longest tool, can greatly reduce the amount of time spent in setting tool lengths, Fig. 6-10-20. The master tool-

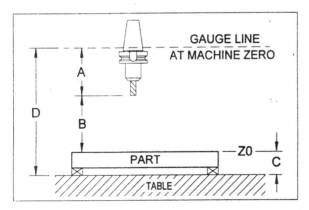

Fig. 6-10-17 The **Z** axis relationship between the machine tool, cutting tool, table top face, and the part height. *(Peter Smid)*

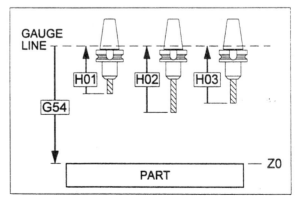

Fig. 6-10-18 The tool length preset away from the machine (tool presetter method). Work offset (G54–G59) must be used. *(Peter Smid)*

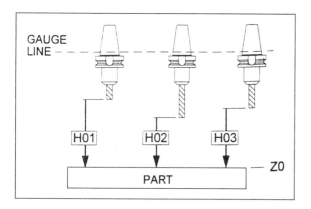

Fig. 6-10-19 The Touch–Off method of setting tool–length offsets. *(Peter Smid)*

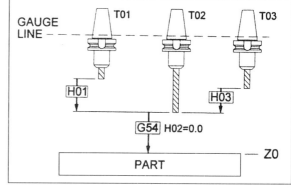

Fig. 6-10-20 Tool–length offsets using the master gage method. *(Peter Smid)*

length compensation method is very efficient if the following procedure is followed:

1. Install the master tool and holder unit into the machine spindle.
2. Zero the **Z** axis and check that the readout screen shows 0.0000.
3. Measure the length of the master tool from the tool tip to **Z0** using the touch method and leave the tool in that position.
4. Instead of registering the measured value to the tool-offset number, register it into the common work offset or one of the G54-G59 work offsets under the **Z** setting; it will be a negative value.
5. While the master tool is still touching the work surface, set the relative **Z** axis readout to zero.
6. Measure all the tools to be used for the job using the touch-off method. The reading will be from the tool tip, not from the machine zero.
7. Enter the measured values under the **H** offset number in the tool length offset screen. It will always be a negative value for any tool shorter than the master tool.

Semi-automatic Tool compensation

Some MCUs are equipped with semi-automatic tool compensation to make allowances for differences in cutting tool lengths. The tool-length compensation feature provides for as many as 64 or more tool length offsets. The programmer uses the same basic tool length for all tools. When the machine is being set up, the operator mounts each tool, starting with the longest, jogs it against a fixed reference stop, and presses a tool-offset button on the MCU. This action records the tool's setting into memory automatically.

All cutting tools for a job are set against the same reference stop, and the difference between each tool's actual length and the basic tool length is entered into memory as a compensation value, Fig. 6-10-21. With this information, the control's computer changes all program steps to take into account the length and diameter of each tool.

CUTTER RADIUS COMPENSATION

Cutter radius compensation, also known as cutter compensation, changes a milling cutter's programmed centerline path to compensate for a small difference (smaller or larger) in cutter radius. On most modern MCUs, the setting is effective for most cuts made using either linear or circular interpolation in the **XY** axis, but does not affect the programmed **Z** axis moves. Usually compensation is in increments of .0001 in. up to +1.0000 in., and most modern controls have a few more offsets available than there are tool pockets in the tool-storage area.

The advantage of the cutter radius compensation feature is that it:

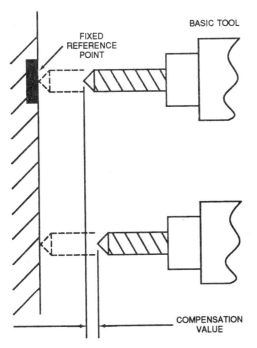

Fig. 6-10-21 Semiautomatic tool–length compensation offsets can be used to quickly set offsets for various length tools. *(Modern Machine Shop)*

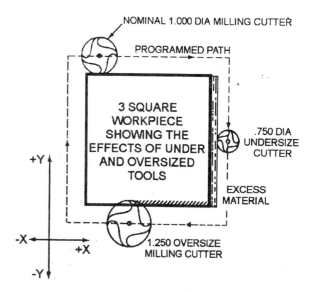

Fig. 6-10-22 Cutter–radius/diameter compensation must be used when machining with under– or over–size cutters. *(Kelmar Associates)*

1. allows the use of cutters that have been sharpened to a smaller diameter.
2. permits the use of a larger or smaller tool already in the machine's storage area.
3. allows the tool to be backed away when roughing cuts are required due to excessive material.
4. permits compensation for unexpected tool or part deflection if the deflection is constant throughout the programmed path.

The basic reference point of the machine tool is never at the cutting edge of a milling cutter, but at some point on its periphery. If a 1.000 in. diameter end mill is used to machine the edges of a workpiece, the programmer would have to keep a .500 in. offset from the work surface in order to cut the edges accurately, Fig. 6-10-22. The .500 in. offset represents the distance from the centerline of the cutter or machine spindle to the edge of the part. Whenever a part is machined, the programmer must calculate an offset path, which is always half the cutter diameter.

Modern MCUs, that have part surface programming, calculate centerline offsets automatically once the diameter of the cutter for each operation is programmed. Many MCUs have operator-entry capabilities that can be used to compensate for differences in cutter diameters; therefore an oversize cutter, or one that has been sharpened, can be used as long as the compensation value for the oversize or undersize cutter is entered.

Programming Cutter-Radius Compensation

The use of cutter radius compensation may vary with different controls. Each control generally has a set of rules that specify how cutter radius compensation is entered and cancelled on that control. The basics of how compensation is programmed are shown in the following example. Be sure to refer to the CNC control manufacturer's manual for the machine control unit being used.

Most controls use three G-codes with cutter radius compensation. G41 is used to activate a cutter-left condition (climb milling with a right-hand cutter). G42 is used to activate a cutter-right condition (conventional milling). G40 is used to cancel cutter-radius compensation. The principle of cutter-radius offset is illustrated in Fig. 6-10-23.

The principle of the tool radius offset

The cutter radius offset makes it possible to program the workpiece dimensions.

Principle:

Tool memory:

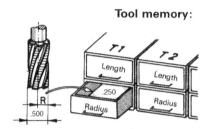

The workpiece dimensions are programmed and the control calculat the tool center path by means of the respective radius value R.

Input into the tool memory:

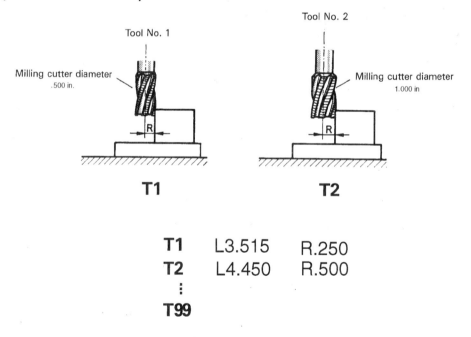

Fig. 6-10-23 **The principle of cutter radius offset.** (*Deckel Maho, Inc.*)

To decide whether to use G41 or G42 code, look in the direction the cutter is moving during machining.

- If the cutter is on the left, use G41.
- If the cutter is on the right, use G42.

Figure 6-10-24 shows some examples that should help to decide when to use the G41, or G42 code. Once cutter-radius compensation is properly activated, the cutter will be kept on the same side of all surfaces until the G40 command is used to cancel the compensation.

CNC Lathe Tooling

The parts produced on a CNC lathe or turning center require only a limited number of cutting tools. Single-point lathe tools are used for some ID (internal diameter) operations and almost all OD (outside diameter) operations such as turning, facing, grooving, etc. Form tools are not commonly used because of the contouring capability of modern MCUs.

Workpieces that are machined with commercial grade tools, tend to vary in size from tool to tool depending on the class of insert. Generally, a class that provides .005 radius on the tool tip would be used for roughing cuts, and a class .0002 would be used for precise finishing cuts. After a looser tolerance insert has been indexed, more time has to be spent in compensating for the variations if the same inserts are used for finishing cuts.

With preset tooling, a variation of insert-type tooling, the toolholder generally has some form of adjustment for length or other characteristic. Tool adjustments to the exact dimensions shown in the part program are made in a special measuring machine or presetting fixture. With presetting, carbide or ceramic inserts can be set to very close tolerances.

Some CNC operators and programmers have accepted qualified tooling because it is a more precise preset system although it is more costly. Machine builders are designing for such tooling, and a specific machine may have slots provided in the turret face to accommodate certain size tools. Two tools of the same designation will cut identically, and the tool will cut the same after each index of the insert.

Tool Nose Radius Compensation

Just as cutter-radius compensation allows the programmer to program milling work surface coordinates, so does tool-nose radius compensation in lathe-type work. Figure 6-10-25 shows why tool-nose radius compensation is used to prevent variations from the work surface on circular and angular cuts.

Usually the nose radius on a lathe tool is small (.015, .031, or .047 in.), so that the variations on the work surface caused by the tool radius will also be small. If the surfaces being machined are very important (angles, chamfers, contours and radii), compensation must be provided for the radius of the tool. Tool-nose radius compensation should only be required on finishing cuts.

The tool-nose radius does not affect the workpiece outline when cutting in the **XZ** direction, Fig 6-10-26A. When cutting in the **Z** direction, machining occurs at the P1 of the cutting tool. However, when the tool starts to cut the taper, Fig. 6-10-26B, the cutting changes to P_2 of the tool nose. Even though the Z_1 dimension was programmed, the taper will not be cut at the correct position because the cutting occurs at different positions on the tool-nose radius. The small difference can be calculated mathematically and the program adjusted accordingly. Modern MCUs do this automatically when the tool information is entered in the program and called into use with the correct code.

To decide whether to use the G41 or G42 command, look in the direction the tool is moving during the cut and note which side of the workpiece the tool is on. For a turning center with tools on the back side of the center line, the tool is on the left, use G41 (e.g. boring toward the chuck); if the tool is on the right, use G42 (turning toward the chuck). Once this is determined, include the proper G code in the tool's first approach to the workpiece. Once tool nose radius compensation is entered, it remains in effect until cancelled by the G40 code.

G41 / G42, G40

To enable the control to calculate the center path of the milling cutter from the program
data and from the tool memory data, it is necessary to communicate to the control where the tool
has to cut.

For this setting there are 3 G-functions:

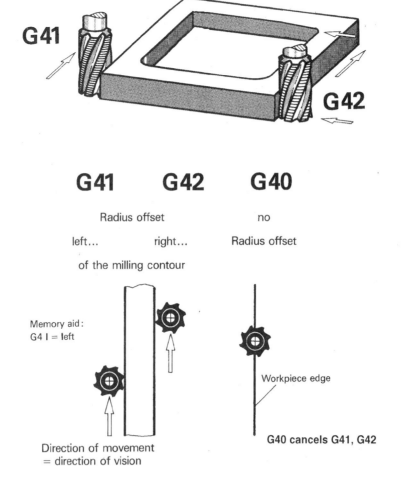

G41	**G42**	**G40**
Radius offset		no
left...	right...	Radius offset
of the milling contour		

Memory aid:
G4 I = left

Direction of movement
= direction of vision

Workpiece edge

G40 cancels G41, G42

Fig. 6-10-24 **Methods of programming cutters for radius compensation.** *(Deckel Maho, Inc.)*

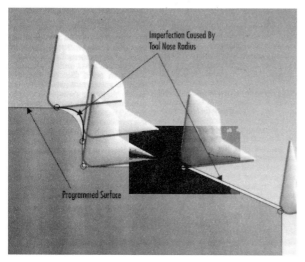

Fig. 6-10-25 Differences in tool–nose radius can cause variations in the surface and accuracy of tapers and contours. (*Modern Machine Shop*)

PROGRAMMING COORDINATE DIMENSIONS

To provide the basics of the programming required to locate edges or hole locations on workpieces, a few examples will be used. It might be well to review the two methods of positioning, absolute and incremental. Absolute positioning is where all dimensions are taken from one fixed point or location. Incremental is where the dimension for the next location is given from the previous location.

Locating points on a workpiece in the coordinate system uses two lines at right angles to each other, one vertical and one horizontal, Fig. 6-10-27. These lines are called axes, and where they intersect is called the origin or zero point. The horizontal line is called the **X** axis; the vertical line is called the **Y** axis.

- Any point to the **right of the Y axis** would be an **X+** (plus) or positive move.
- Any point to the **left of the Y axis** would be an **X–** (minus) or negative move.
- Any point **above the X axis** would be a **Y+** move.
- Any point **below the X axis** would be a **Y–** move.

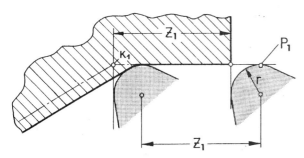

Fig. 6-10-26A The final form is produced by P_1 on the tool radius, therefore the full length programmed will not be cut. (*Modern Machine Shop*)

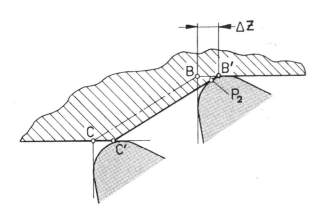

Fig. 6-10-26B When taper turning, P_2 of the nose radius produces the final taper outline. (*Modern Machine Shop*)

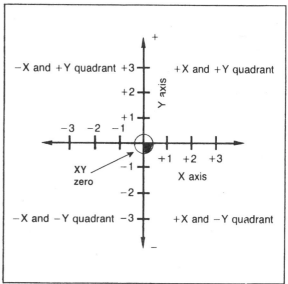

Fig. 6-10-27 The intersecting lines, at right angles to each other, form the four quadrants of a coordinate system. (*Allen-Bradley*)

Drilling Operations

The hole locations on the absolute example shown in Fig. 6-10-28 will be programmed, assuming that the starting point for the center of the tool or machine spindle is located over the **XY** zero at the top left-hand corner of the part. After locating at hole #3, return the tool to **XY** zero.

Hole #1 = X1.000 (positive move right of **XY** zero)

= Y–.750 (negative move below **XY** zero)

Hole #2 = X2.000 (2.000 right of **XY** zero)

= Y–.750 (generally does not have to be programmed since **Y** location is same as hole #1)

Hole #3 = X3.000 (3.000 right of **XY** zero)

XY Zero = X0 Y0 (tool returns to start position)

In the incremental example, Fig. 6-10-29, the **XY** zero is at the bottom left-hand corner of the part. Program the center location of all holes in numerical sequence, and after hole #3, return the tool to the **XY** zero position.

Hole #1 = X1.000 (positive move right of **XY** zero)

= Y–.750 (positive move above **XY** zero)

Hole #2 = X1.000 (positive move right of hole #1)

= Y0 (indicates no change **Y** location)

Hole #3 = X1.000 (positive move right of hole #2)

= Y0 (no change)

XY Zero = X–3.000 (negative move from hole #3)

Y.750 (negative move from hole #3)

Milling Operations

In milling operations, a point on the circumference (periphery) of the cutter does the machining. The cutter path is not the contour or outer surface, but a distance equal to the radius of the cutter away from the part. For example, if a .750 in. diameter end mill is being used to machine the edges of the parts shown in Figs. 6-10-28 and 6-10-29, the center of the machine spindle (or cutter) must be programmed (offset) .375 in. away from all edges.

To program for machining the edges of the absolute part shown in Fig. 6-10-28, the start point will be the **XY** zero at the top left-hand corner of the part, plus about .050 in. for safety. The machining is to be in a clockwise (climb milling) direction around the edges of the part.

1. The .750 in. diameter cutter must be positioned .375 in. to the left of **XY** zero, or in a minus direction, and .375 in. above the **XY** zero in a plus direction and .050 for clearance.
   - **Start position** = X–.375 Y.375

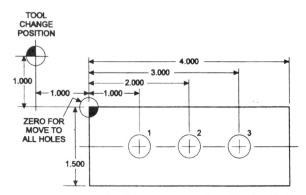

Fig. 6-10-28 In absolute programming (G90), all dimensions must be taken from the XY zero at the top left–hand corner of the part. *(Kelmar Associates)*

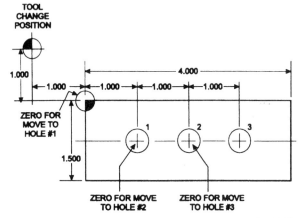

Fig. 6-10-29 In incremental programming (G91), all dimensions are taken from the previous point. *(Kelmar Associates)*

2. **Milling the top edge**
   - X4.375 (the –.375 from left of part to .375 past right-hand edge)
   - Y.375 (remains same, and on most controls, does not have to be programmed again)
3. **Milling the right-hand edge**
   - X4.375 (does not have to be programmed; no move in **X** direction)
   - Y–1.875 (cutter center moves .375 in. below bottom edge)
4. **Milling the bottom edge**
   - X–.375 (cutter center moves .375 in. past left-hand edge of part)
   - Y–1.875 (position stays at –1.875 in.)
5. **Milling the left-hand edge**
   - X–.375 (cutter remains at –.375 in. or start point)
   - Y.375 (cutter returns to start point)

To program for incremental machining of the edges of the part shown in Fig. 6-10-29, the start point will be the **XY** zero at the bottom left-hand corner of the part. The machining, with a .750 in. diameter cutter, is in a clockwise (climb milling) direction around the edges of the part.

1. The .750 in. diameter cutter must be positioned .375 in. to the left of the **XY** zero, or in a minus direction, and .375 in. below the bottom edge, or a minus direction.
   - **Start position** = X–.375, Y.375
2. **Milling the top edge**
   - X4.750 (cutter moves .375 in. past right-hand edge)
   - Y0 (cutter remains at .375 in. above part)
3. **Milling the right-hand edge**
   - X0 (cutter remains at .375 in. past right-hand edge)
   - Y–2.250 (cutter moves .375 in. below bottom edge)
4. **Milling the bottom edge**
   - X–4.750 cutter returns to start point)

- Y0 (cutter remains at .375 in. below bottom edge)
5. **Milling the left-hand edge**
   - X0 (cutter remains at –.375 in., or start point)
   - Y2.250 (cutter moves .375 in. above top edge)

SUMMARY

- The three important factors that must be considered in CNC programming are the machine tool, the workpiece, and the cutting tool.
- Machine and work reference points are important to CNC programming. The machine reference point is the start of the machine's coordinate system; the work reference point is the start of the part programming.
- Work coordinates are used as reference points when locating workpieces and fixtures. Machine coordinates are often used as tool-change positions.
- Machine zero is a fixed position on a CNC machine that can be reached at any time through the MCU, MDI, or program code.
- The location of the workpiece reference point should be chosen so that it is easy to convert drawing dimensions to coordinates, best for setup, accuracy, safety, and machining operations.
- Before machining a part on a CNC machining center, the spindle (cutting tool) should be aligned with the **XY** axes of the part edges and the **Z** axis with the surface of the part.
- Offsets or compensation are used to make adjustments in the program for tool length, radius, nose radius, cutter resharpening, deflection, and adjustment of tapers.
- Presetting tools, touch-off, master tool, and semi-automatic tool compensation methods can be used to compensate for variations in tool length.
- Tool-nose radius on turning tools must be considered during the programming of tapers and contour forms.

KNOWLEDGE REVIEW

1. What common factor determines the accuracy and locations on a part?

Machine and Work Coordinates

2. Define the purpose of the (a) absolute position register, (b) offset register.

Reference Points

3. List the three main reference points used on a CNC vertical mill for programming.

Machine Reference Point

4. Define machine zero.
5. Where is the machine reference (zero) point usually found on:
   (a) CNC vertical machining center?
   (b) CNC turning center?

Workpiece Reference Point

6. Where is the workpiece reference point usually located on:
   (a) machining center?
   (b) turning center?

Tool Settings and Offsets

7. In point form, explain how to set the X and Y axes to zero with an edge finder.
8. Briefly explain how to set the Z axis to zero.

Work Settings and Offsets

9. Where is the R-plane generally set in relation to the work surface?

Tool Offsets

10. Why is it important to compensate for differences in tool diameter and length?
11. Name two methods of setting for the differences in tool length.
12. What is the purpose of the gage line?

Tool Length Compensation

13. Name four methods of setting tool-length compensation.
14. Explain how semi-automatic tool compensation works.

Cutter Diameter Compensation

15. State three advantages of cutter diameter compensation.
16. State the purpose of the following cutter radius compensation codes: G41, G42, G40.
17. What effect does tool nose radius have on angular and contour surfaces?

Programming Coordinate Dimensions

18. Define absolute and incremental positioning.
19. In relation to the coordinate system, locate the following:
    (a) a point to the right of the Y axis and below the X axis.
    (b) a point to the left of the Y axis and above the X axis.

SECTION 7

CNC Machining Centers

CNC MACHINING CENTERS

CNC machining centers evolved from the need to be able to perform a variety of different operations and machining sequences on a workpiece on a single machine in one setup. In the past, many parts that required machining on several machines could spend weeks on the shop floor waiting and moving from machine to machine. A workpiece may spend only 5% of its time in the shop on a machine, and only about 30% of that 5%, or 1.5%, in actual machining time greatly reducing manufacturing productivity, Fig. 7-1. Operations such as milling, contouring, drilling, counterboring, boring, spotfacing, and tapping can be performed on CNC machining centers in any sequence and require only one setup. Machining centers equipped with automatic-tool changers, rotary tables, and rotary work heads make the maching center a very versatile machine while reducing the operator intervention during the cutting cycle.

Since the introduction of the first NC controlled machining centers in the late 1950s they have developed over the years into the highly-productive CNC machining centers of today. A major factor in this development was the introduction of computer controls in the 1970s, which resulted in better positioning systems that in turn produced more accurate parts. Continual development in sturdier and more versatile machine tools, software programs, cutting tool materials, tooling systems, and material-handling systems have made the CNC machining center one of the most important and widely used machine tools in the world.

The machining centers of today, Fig. 7-2, are flexible enough to handle a wide variety of workpieces using multiple-axis control to manufacture parts having simple to complex forms to a high degree of accuracy. These machines, using hundreds of different cutting tools, are cost effective on a single part or a large production run and can perform operations such as milling, drilling, tapping, boring, and other mill-related operations. CAD/CAM and Virtual Reality software have made it possible to simulate machining conditions on a computer screen to prove the accuracy of the

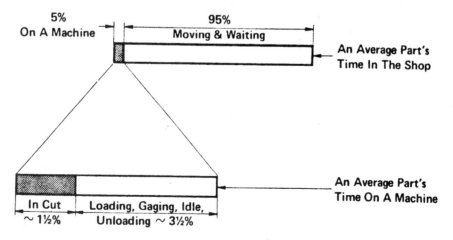

Fig. 7–1. A distribution of time a part spends in a conventional shop shows only a small percentage of time is actually spent machining the part. *(Cincinnati Machine, A UNOVA Co.)*

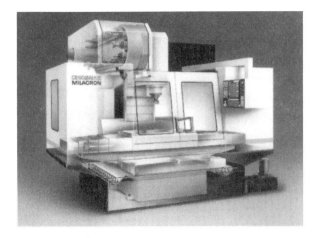

Fig. 7-2. Modern machining centers can perform a wide variety of machining operations quickly and accurately. *(Cincinnati Machine, A UNOVA Co.)*

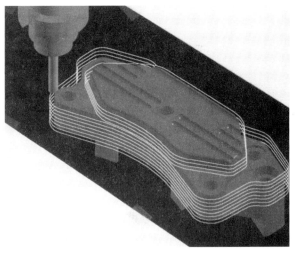

7-3. The graphic simulation of the cutter path for machining a part can indicate machine crashes or machining errors. *(Haas Automation, Inc.)*

CNC program before machines, tools, and fixtures are purchased, Fig. 7-3. This process allows a CNC program to be checked, and corrected before running the part on a machine, avoiding costly crashes, broken tools, and scrap parts.

Types of Machining Centers

There are three main types of CNC machining centers: the vertical spindle, the horizontal spindle, and the universal machining centers. Many of these machines are available in bench-top teaching models, Fig. 7-4, that are ideal for teaching purposes since they use the same basic electronic controls as do the larger industrial machines. All of these CNC machining centers will be covered in more detail in Units 12 to 15.

7-4. CNC bench-top mills are ideal for teaching the principles of programming. *(Denford, Inc.)*

Unit 11

Mill Tooling and Workholding Systems

Modern CNC machines and manufacturing processes require high-performance toolholders, cutting tools, and workholding systems to consistently produce high-quality parts. Regardless of how well a machine is designed and built or how good the cutting tool used, the success of any machining operation depends upon whether the part is located accurately and held securely.

Modern modular tooling systems are precision tools that consist of an adapter (shank) to fit into the machine spindle, a reduction or extension adaptor, and the cutting toolholder, Fig. 7-11-1. The precision and reliability built into modular tooling systems provide maximum rigidity with very strong face-to-face clamping forces to reduce or eliminate vibration.

The most important function of any workholding device is to hold the part so that the surface to be machined is in the correct relationship to other surfaces as indicated on the part drawing. The part must be held securely enough that it can withstand the forces created during the machining operation without becoming loose or moving.

After completing this unit, you should be able to select:

1. The proper and most efficient cutting tools for the machining operations required on the part.
2. The modular tooling system that provides the most flexibility and allows for fast tool changes.
3. The proper workholding system to hold each part accurately and securely.

KEY TERMS

adaptive control	sequential tooling
modular tooling	toolchangers
preset tools	tool identification
random tooling	workholding systems

SPEEDS AND FEEDS

The most important factors affecting the efficiency of a CNC machining center are cutter speed, feed, and depth of cut. **Speed** is the rate that a cutting tool revolves generally measured in feet or meters per minute. If a cutting tool is run too slowly, valuable time will be wasted, resulting in lost production. Too high a speed will create too much heat and friction at the cutting edge of the tool, which quickly dulls the cutter, Fig. 7-11-2. The machine would then have to be stopped to either

Fig. 7-11-1 Modular tools are precision tooling systems that allow cutting tools to be replaced quickly and accurately. *(KPT Kaiser Precision Tooling, Inc.)*

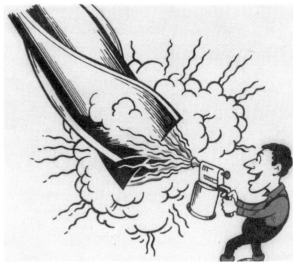

Fig. 7-11-2 High speed creates excessive heat during machining that quickly dulls cutters and shortens their life. (*Weldon Tool Co.)*

recondition or replace the cutter. Somewhere between these two extremes is the efficient cutting speed for each material.

Feed is the rate at which the work is fed into a revolving cutter. If the work is fed too slowly, time will be wasted, resulting in lost production, and cutter chatter, which shortens the life of the cutter. If work is fed too fast, the cutter teeth can be broken, Fig. 7-11-3. Much time will be wasted if several shallow cuts are taken instead of one deep cut or roughing cut. Therefore, speed, feed, and depth of cut are three important factors that affect the life of the cutting tool and the productivity of the machine tool. Charts and tables containing speeds, feeds, and depth of cut are available from cutting tool manufacturers and Machinery's Handbook.

The programmer should select the proper speeds and feeds for each part to be machined, so that the part is produced in the shortest period of time with good cutting tool life. Generally, cutting-tool breakage occurs because the

cutter is dull, or the depth of cut is changed because of variances in workpiece thickness. Whenever a cutting tool becomes dull or is broken, the CNC machine must be stopped to recondition or replace the cutting tool. To get the best productivity from a CNC machine tool, the selected speeds and feeds should be those that give the best productivity with the longest tool life.

Fig. 7-11-3 Too fast (heavy) a feed can cause the cutting edges to chip or the cutter to break. *(Weldon Tool Co.)*

CUTTING TOOLS

The selection of the proper cutting tools for each operation on a machining center is essential to producing an accurate part. Generally not enough thought and planning are given to the selection of cutting tools for each particular job. The CNC programmer must have a good knowledge of cutting tools and their applications in order to properly program any part.

Machining centers use a variety of cutting tools to perform various machining operations, Fig. 7-11-4. These tools may be conventional high-speed steel, cemented carbide inserts, CBN (cubic boron nitride) inserts, or polycrystalline diamond insert tools. Some of the common tools used are end mills, drills, taps, reamers, boring tools, etc. Studies show that machining center time consists of 20 percent milling, 10 percent boring, and 70 percent hole-making in an average machine cycle. On conventional milling machines, the cutting tool cuts approximately 20 percent of the time, and on machining centers the cutting time can be as high as 75 percent. The end result is that there is a larger consumption of disposable tools through increased tool usage.

End mills

End mills and shell end mills, Fig. 7-11-5, are widely used on machining centers. They are capable of performing a variety of machining operations such as face, pocket, and contour milling, spotfacing, counterboring, roughing and finishing of holes using circular interpolation.

Drills

Conventional as well as special drills are used to produce holes, Fig. 7-11-6. Always choose the shortest drill that will produce a hole of the required depth. As drill diameter and length increase, so does the error in hole size and location. Stub drills are recommended for drilling on machining centers.

Center drills

Center drills, Fig. 7-11-7, are used to provide an accurate hole location for the drill that is to follow. The disadvantage of using center drills is

Fig. 7-11-4 CNC uses a wide variety of cutting tools, each designed to perform a specific machining operation. *(KPT Kaiser Precision Tooling, Inc.)*

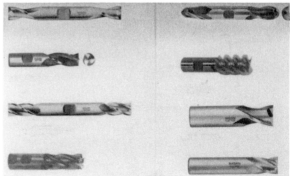

Fig. 7-11-5 End mills and shell end mills are widely used on machining centers. *(Niagara Cutter)*

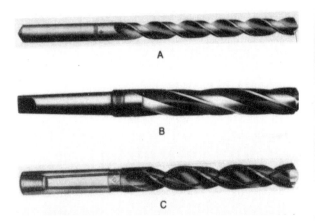

Fig. 7-11-6 Many types of drills are used for producing holes: (A) high-helix drill, (B) core drill, (C) oil-hole drill. *(Cleveland Twist Drill Co.)*

that the small pilot drill can break easily unless care is used. An alternative to the center drill is the spotting tool, which has a 90° included angle and is widely used for spotting hole locations.

Taps

Machine taps, Fig. 7-11-8, are designed to withstand the torque required to thread a hole and clear out the chips. Tapping is one of the most difficult machining operations to perform because of the following factors:

- Inadequate chip clearance
- Inadequate supply of cutting fluid
- Coarse and fine threads in various materials
- Speed and feed of threading operations are governed by the lead of the thread
- Depth of thread required

Fig. 7-11-7 Center drills are commonly used to accurately spot the location of holes. *(Cleveland Twist Drill Co.)*

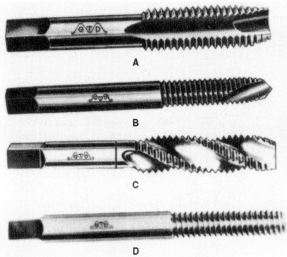

Fig. 7-11-8 Taps are used to produce internal threads: (A) gun, (B) stub flute, (c) spiral flute, (D) fluteless. *(Greenfield Tap and Die)*

Reamers

Reamers are available in a variety of designs and sizes, Fig. 7-11-9. A reamer is a rotary end cutting tool used to accurately size and produce a good surface finish in a hole that has been previously drilled or bored.

Boring tools

Boring is the operation of enlarging a previously drilled, bored, or cored hole to an accurate size and location with a desired surface finish. This operation is generally performed with a single-point boring tool, Fig. 7-11-10. When a boring bar is selected, the length and diameter should be carefully considered: as the ratio between length

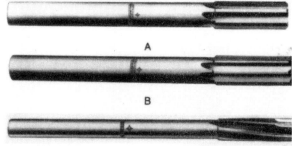

Fig. 7-11-9 Reamers are used to size a hole and produce a good surface finish. *(Cleveland Twist Drill Co.)*

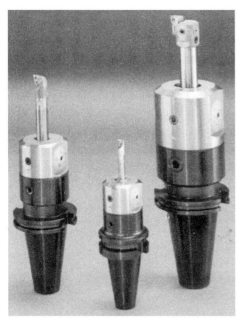

Fig. 7-11-10 Single-point boring tools are used to enlarge a hole at the required location. *(KPT Kaiser Precision Tooling, Inc.)*

and diameter increases, the rigidity of the boring bar decreases. For example, a boring bar with a 1:1 length-to-diameter ratio is 64 times more rigid than one with 4:1 ratio.

TOOLING SYSTEMS

Toolholders

The machining center, a multifunction machine tool, uses a wide variety of cutting tools such as drills, taps, reamers, end mills, face mills, boring tools, etc., to perform various machining operations on a workpiece, Fig. 7-11-11. For these cutting tools to be inserted into the machine spindle quickly and accurately, they must have toolholders with the same taper shank to suit the machine spindle. The most common taper used in CNC machining center spindles is the No. 50 taper, which is a self-releasing taper. The toolholder must also have a flange or collar for the tool-change arm to grab, and a stud, tapped hole, or some other device for holding the tool securely in the spindle by a power drawbar or other holding mechanism, Fig. 7-11-12.

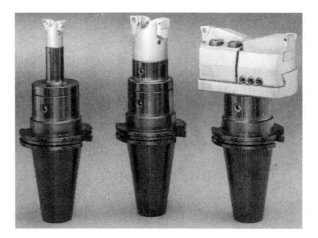

Fig. 7-11-11 Standard toolholders can be fitted with a variety of cutting-tool components for specific machining operations. *(KPT Kaiser Precision Tooling, Inc.)*

When preparing for a machining sequence, the tool assembly drawing is used to select all the cutting tools required to machine the part. Each cutting tool is then assembled off-line in a suitable toolholder and preset to the correct length or diameter. Once all the cutting tools are assembled and preset, they are loaded into specific pocket locations in the machine's tool-storage magazine where they are automatically selected as required by the CNC part program.

Tool Identification

CNC machine tools use a variety of methods to identify the various cutting tools that are used for machining operations. The most common methods of identifying tools are:

1. **Tool pocket locations**
Tools for early machining centers were assigned a specific pocket location in the tool-storage magazine, and each tool was called up for use by the part program.
2. **Coded rings on toolholders**
A special interchange device reader is used to identify some tools by special coded rings on the toolholder.
3. **Tool assembly number**
Most modern MCUs have a tool identification feature that allows the part program to recall a tool from the tool-storage magazine pocket by using a five- to eight-digit tool assembly number.

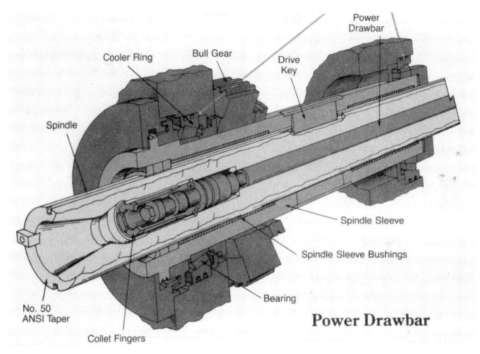

Fig. 7-11-12 The power drawbar is a hydraulically-operated tool that locks adaptors in place quickly and securely. (*Giddings and Lewis, Inc.*)

Each tool assembly number may be assigned a specific pocket in the tool-storage magazine by the tool data program, by the operator using the MCU, or by a remote tool management console.

Tool Management Program

To achieve the best productivity from any machine tool, it is important to have a tool management program that covers all aspects of cutting tools. No CNC machine can reach the best productivity potential unless the best cutting tools and corresponding toolholders are available when they are required. A good tool-management program should include tool design, standard coding system, purchasing, good tooling practices, part programming which is cost effective, and the best use of cutting tools on the machine.

Good cutting-tool management programs include the following:

1. **Standard policy**
   - A standard policy regarding cutting tools that everyone can understand.
   - The role that each person has in selecting the proper cutting tools must be clearly defined.

2. **Cutting tool dimensional standards**
   - All cutting tools purchased or specially made must conform to established cutting tool dimensional standards.
   - When cutting tools are resharpened, they should be ground to the next CNC standard.
   - The part programmer must use standard cutting tools for programming purposes.

4. **Rigid cutting tools**
   - Always use the shortest cutting tool possible to avoid chatter and deflection.
   - Solid cutting toolholders or those specifically designed to eliminate vibration should be used to provide rigidity.

5. **Tool preparation**
   - There must be a set policy on tool setting, compensation, and regrinding that is understood by everyone.

6. **Indexable insert tools**
   - Cemented carbide insert tooling generally provides good wear resistance, high productivity, and dimensional accuracy.
   - CBN (cubic boron nitride) inserts should be used on hard ferrous metals where cemented carbides break down.
   - PCD (polycrystalline diamond) inserts should be used for machining abrasive nonferrous materials.

The effectiveness of a tool-management program depends mostly on the part programmer. The programmer must have a good knowledge of machining processes and the cutting tool required for each operation. Modern MCUs (machine control units) have features that can make tool-management programs effective.

RANDOM AND FIXED (SEQUENTIAL) TOOLING SYSTEMS

Most modern CNC machine tools are equipped with automatic tool changers to quickly replace cutting tools for the next machining operation. These tool changers may be designed for either random or fixed tooling selection.

Random tool selection, the most common on CNC machining centers, is a system where there is no specific pattern of tool selection. Random tooling is generally in wider use in industry because of the flexibility it offers over sequential tooling.

- Each tool is given a specific tool identification number and is loaded into a specific pocket in the tool magazine. As each tool is required for use in the CNC program, the previous tool is removed from the machine tool spindle by the tool changer arm and replaced in the correct tool-magazine pocket.
- The new tool, selected by the CNC program, is taken from the correct tool-magazine pocket and inserted into the machine spindle. Whenever a certain cutting tool, or one that has been used before, is needed for machining, the MCU knows where to find it.

Fixed (Sequential) tool selection is a system, used on older and less expensive CNC machin-

ing centers, where tools must be loaded in the sequence in which they will be used during machining a part. Therefore it is important that the correct sequence of tooling be programmed and loaded in the tool magazine as they are required to complete the machining operations on a part.

- If the cutting tools are not in the correct order, the next tool is automatically selected and the machine may try to tap a hole with an end mill. Therefore, when it is necessary to use a tool more than once, it is necessary to load similar tools in the tool magazine in the order that they are to be used.

ADAPTIVE CONTROL

A feature that is becoming very popular is **torque control machining,** where the torque of the machining operation is calculated from measurements at the spindle drive motor, Fig. 7-11-13. This device can increase productivity by preventing or sensing potential damage to the cutting tool because of dulling. The torque is measured when the machine is turning but not cutting, and this value is stored in the computer memory. As

Fig. 7-11-13 Torque control increases or decreases the feed rate, depending on the cut or the dullness of the cutting tool during a machining cycle. *(Cincinnati Machine, A UNOVA Co.)*

the machining operation begins, the stored value is subtracted from the torque reading at the motor. This calculation gives the net cutting torque, which is compared with the programmed torque or limits stored in the computer program. If the net cutting torque exceeds the programmed torque limits, the computer will act by reducing the feed rate, turning on the coolant, or even stopping the cycle.

The feed rate will be lowered whenever the horsepower requirements exceed the rated motor capacity or the programmed code value. The system display of three yellow lights advises the operator of the operational conditions in the machine at the time. A left-hand yellow light indicates that the torque control unit is in operation. The middle yellow light indicates that the horsepower limits are being exceeded. The right-hand light comes on when the feed rate drops below 60% of the programmed rate.

The meter, Fig. 7-11-13, indicates the cutting torque (or operational feed rate) as a percent of the programmed feed rate. As the torque increases, the machine will reduce the feed. The problem might be caused by excessive material on the workpiece, or a tool might be very dull or broken. If the tool is dull, the machine will finish the operation and a new backup tool of the same size will be selected from the storage chain when that operation is performed again. If the torque is too great, the machine will stop the operation on the workpiece and program the next piece into position for machining.

MODULAR TOOLING SYSTEMS

Modern modular tooling systems are precision tools that consist of an adapter (shank) to fit into the machine spindle, a reduction or extension adaptor, and the cutting toolholder, Fig. 7-11-14. The precision and reliability built into modular tooling systems provide maximum rigidity with very strong face-to-face clamping forces to reduce or eliminate vibration.

Modular tooling, sometimes called quick-change tooling, allows cutting tools to be re-

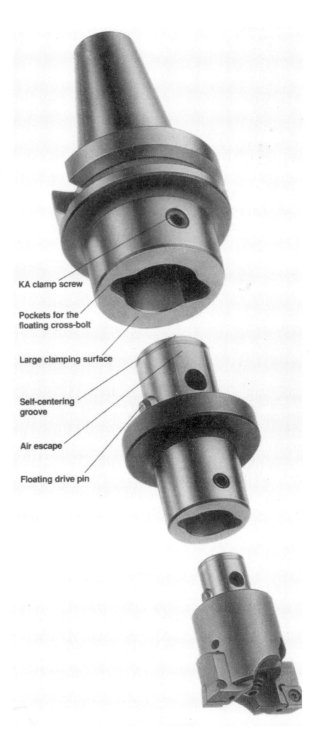

KA clamp screw

Pockets for the floating cross-bolt

Large clamping surface

Self-centering groove

Air escape

Floating drive pin

Fig. 7-11-14 Modular tools offer a precise, rigid assembly of various tool components. (*KPT Kaiser Precision Tooling, Inc.*)

placed quickly and accurately resulting in increased machining time. A few of the many advantages of modular tooling systems are:

1. The system allows the accurate presetting of tools for an entire job, reducing machine setup time.
2. New tools can be provided quickly by exchanging components in stock.
3. Quick, positive change of tools reduces the amount of machine downtime and increases productivity.
4. Modular tools are as rigid as solid tooling and have a repeatability accuracy of .0001 in.
5. Tooling costs are reduced because of the savings in tool-change time, tool-setup time, time required for trial cuts, and reduced scrap.

WORKHOLDING DEVICES

It is important to ensure that a workpiece setup is safe. The workpiece must be securely fastened, and the setup must be rigid enough to withstand the forces that will be present during the machining operation. If the workpiece or the holding device becomes loose during machining, damage can result to the tooling and/or the machine.

The machine operator should be sure that all workholding devices are free from chips and burrs before use. The workholding devices, generally specified by the programmer, should be located in the proper position on the machine table. Failure to follow these instructions may result in operator injury, damage to the machine, or scrap workpieces.

Types of Workholding Devices

The most important function of any workholding device is to hold the part so that the surface to be machined is in the correct relationship to other surfaces as indicated on the part drawing. The part must be held securely enough that it can

withstand the forces created during the machining operation without becoming loose or moving. Although workholding devices differ due to the shape and size of the part, the most commonly used are:

The **Swivel-base vise,** Fig. 7-11-15A, which may be bolted to the machine table or a subplate. The swivel base enables the vise to be swivelled 360° in a horizontal plane.

Angle plates, Fig. 7-11-15B, are L-shaped pieces of cast iron or steel accurately machined to a 90° angle. They are made in a variety of sizes and have holes or slots that provide a means for fastening the workpiece.

V blocks, Fig. 7-11-15C are generally used in pairs to support cylindrical work. A U-shaped clamp may be used to fasten the work in a V block.

Step blocks, Fig. 7-11-16A, are used to provide support for strap clamps when work is being fastened to the table or workholding device.

Clamps or Straps, Fig. 7-11-16B are used to fasten work to the table, angle plate, or fixture. They are made in a variety of sizes and are usually supported at the end by a step block and bolted to the table by a T bolt. It is good practice to place the T-bolt in the clamp or strap as close to the work as possible.

Support jacks, Fig. 7-11-16C, are used to support the workpiece to prevent distortion of the workpiece during clamping.

Parallels, Fig. 9-11-16D, are flat, square, or rectangular pieces of metal used to support the workpiece for setup.

Subplates are generally flat plates that may be fitted to the machine table to provide quick and accurate location of workpieces, workholding devices, or fixtures. The fixturing holes in these sub plates are accurately located and, when set up on the machine table in relation to the machine datum, provide the programmer with known locating positions.

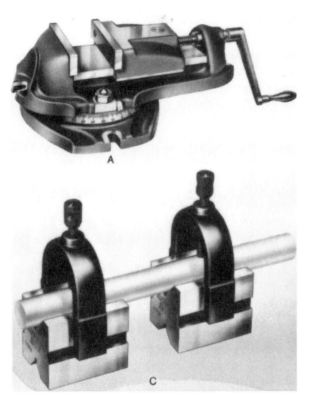

Fig. 7-11-15 (A) The swivel-base vise can be swiveled through 360° in a horizontal plane; (B) Angle plates are machined to an accurate 90° angle *(Kelmar Associates)*; (C) V-blocks are used to hold round work for machining *(L. S. Starrett Co.)*.

Fig. 7-11-16   (A) Step blocks support the end of the clamp *(Northwestern Tools, Inc.)*; (B) Clamps and straps are used to fasten work to the machine table *(J. W Williams & Co.)*; (C) Support jacks are used to support and prevent the work from distorting when being clamped *(Kelmar Associates)*; (D) Parallels are used to support the workpiece.

FIXTURES

CNC eliminates many of the expensive jigs and fixtures that were previously necessary to hold and locate a workpiece on conventional machine tools. The repetitive position accuracy of an CNC machine tool also eliminates the need for guide bushings, which were previously required to locate the cutting tool.

CNC fixtures, Fig. 7-11-17, are used to accurately locate a part and hold it securely for machining operations. Fixture design should be kept simple so that the time required to load and unload a part is kept as short as possible. Since this is nonproductive time, the savings here will result in corresponding savings in the cost of producing a part. When designing a fixture to hold a part, it is important to consider the following points:

1. **Positive location**: The fixture must hold a workpiece securely enough to prevent the workpiece from linear movement in the X,

Y, and Z axes, and rotational movement in either direction about each axis.

2. **Repeatability**: Identical parts should always be held in exactly the same location for every part change.

3. **Ruggedness**: Fixtures must be designed to withstand the shock created during the machining and loading/unloading cycles.

4. **Rigidity**: The workpiece must be held securely to prevent any movement due to the forces created by the machining operation.

5. **Design**: Modular fixtures using standard components are quicker to produce and less costly than custom fixtures. They can also be quickly modified to accommodate differently shaped parts, Fig. 7-11-18.

6. **Low profile**: Parts of the fixture or the necessary clamping devices to hold the part should be designed to allow free movement for the cutting tool at any point in the machining cycle.

7. **Part loading/unloading:** The fixture and its clamping devices should be designed so that they do not interfere with the rapid loading or unloading of a part.

8. **Part distortion:** The fixture should be designed so that the part being machined is not distorted by gravity, machining forces, or clamping forces. Stress should never be put on a part by the clamping forces; otherwise the machined part will distort when the clamping forces are removed.

Fig. 7-11-17 Various parts can be clamped to a tombstone fixture ready for the machining center. *(Prohold Workholding, Inc.)*

Modular Fixtures

Modular fixtures provide many of the advantages of permanent fixtures but are flexible enough to accommodate various shapes of workpieces by changing certain components. A modular fixture can be built from a set of standard components to hold a certain part shape, Fig. 7-11-18. After the production run is complete, the fixture can be disassembled to allow the components to be reused. A manufacturer can thus make fixtures at any time to suit the part to be manufactured.

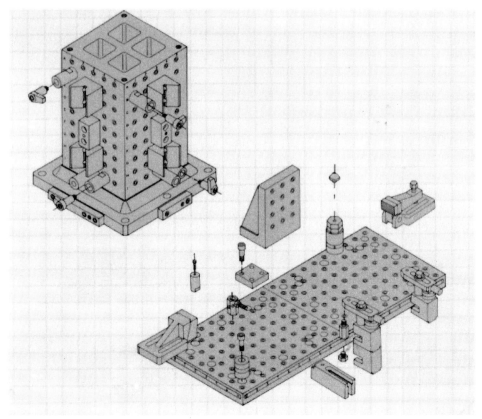

Fig. 7-11-18 Modular workholding fixtures combine ideas and elements of both permanent and temporary work-holding devices to make inexpensive fixtures. *(Carr Lane Manufacturing Co.)*

Clamping Hints

1. Always place the bolt as close to the work as possible.
2. Place a piece of soft metal ("packing") between the clamp and the workpiece to prevent damage to the workpiece and to spread the clamping force over a wider area.
3. Make sure the packing does not extend into the machining path of the cutting tool.
4. Use the table slots to prevent round work from moving.
5. Use two clamps whenever possible.
6. Parts that do not lie flat should be shimmed to prevent the work from rocking. Shimming will also prevent distortion when the work is clamped.
7. Tighten clamping bolts evenly to prevent workpiece distortion.

SUMMARY

- Speeds, feeds, and depth of cut are important factors that affect the life of a cutting tool and the productivity of any machining operation.
- A CNC programmer should have a good knowledge of cutting tools and machining processes to properly program a part.
- The most common cutting tools used on CNC machining centers are end mills, drills, taps, reamers, and boring bars.
- CNC toolholders can be identified by the tool pocket location, coded rings, and tool assembly number.
- Two types of cutting tool selection programs used on CNC machining centers are the random and fixed types.
- Modular tooling systems are widely accepted because of their fast and accurate tool changes,

reduced tooling costs, and increased productivity.
- Effective workholding systems are those that position the part accurately and hold it securely during the forces exerted by the machining operation.
- The most common workholding devices used

on CNC machining centers are vises, angle plates, clamps and straps, subplates, tombstones, and fixtures.
- An effective fixture design provides positive part locations, rigidity, easy part loading/unloading, and no part distortion.

KNOWLEDGE REVIEW

1. Name three qualities that modern CNC machines require to produce high-quality parts consistently.

Speeds and Feeds

2. What effect does (a) high speed, and (b) low speed have on the cutting tool life and productivity?
3. Name the two factors that should be considered to get the best productivity from a CNC machine tool.

Cutting Tools

4. What knowledge should a CNC programmer have about tools and applications in order to properly program a part?
5. List six machining operations commonly performed by end mills.
6. Why are center drills important when locating the position of holes?
7. For what purpose are reamers used?

Tooling Systems

8. Name three methods of identifying cutting tools used on CNC machining centers.
9. List the five factors important to a good cutting-tool management program.
10. Define the difference between random and fixed (sequential) tooling systems.

Modular Tooling Systems

11. What are five advantages of modular tooling systems?

Workholding Devices

12. Name two important functions of a good workholding device.
13. For what purpose are the following used? (a) V-blocks (b) clamps or straps (c) subplates
14. Name five of the most important features that a good fixture should have.

Unit 12
CNC Mill

The first NC machines, introduced in the late 1950s, were drilling machines that were only capable of point-to-point positioning (straight-line motions). Numerical Control (NC) quickly developed into Computer Numerical Control (CNC), and CNC machine tools, because of their many advantages, became widely accepted in the world so that by the mid 1990s, 90% of the machine tools manufactured were CNC controlled and only 10% were manual machines.

Machining centers evolved from the need to be able to perform a variety of operations and machining sequences on a workpiece on a single machine in one setup. Many parts require machining on several machines and may spend weeks on the shop floor waiting and moving from machine to machine. Operations such as milling, contouring, drilling, counterboring, boring, spotfacing, and tapping can now be performed on machining centers in any sequence and require only one setup. Machining centers equipped with automatic tool changers, rotary tables, and rotary work heads give versatility and reduce the need for operator intervention during the cutting cycle.

OBJECTIVES

After completing this unit, you should be able to:

1. Understand the main types of CNC machining centers and the machining operations for which each is designed.
2. State eight main advantages of machining centers.
3. Know the direction of movement when positive (+) and negative (−) moves are made in the **X, Y, Z** axes.

KEY TERMS

fixed column	positive moves
negative moves	quadrants
origin point	servo system
point-to-point positioning	Variax

TYPES OF MACHINING CENTERS

There are four main types of machining centers: the vertical machining center, the horizontal machining center, the universal machining center, and the VARIAX machine.

• Vertical machining centers, Fig. 7-12-1, have the cutting tool held in a vertical position. They are generally used to perform operations on flat parts that require motion in the X, Y, and Z axes.

Horizontal Spindle Type

• Horizontal machining centers have the cutting tool held in a horizontal position. They are available in two types, the **travelling-column**, and the **fixed-column** type, Fig. 7-12-2. They can machine parts on more than one side in one clamping, and find wide use in flexible manufacturing systems.

• Universal machining centers, Fig. 7-12-3, allow the spindle to be programmed for either vertical or horizontal machining at any time during the program. This design combines the features of both the vertical and horizontal machining centers, allowing the machining of all sides of a part in one setup.

• VARIAX™—The Machine Tool of the Future,

Fig. 7-12-2 A horizontal machining center is used to machine the side surfaces of a part. (*Giddings & Lewis, Inc.*)

developed by the Giddings & Lewis Corp., is a radical departure from the conventional design of machining centers, Fig. 7-12-4. The advanced technological design of the machine uses some of the most basic physical laws of nature. The triangles formed by its six legs, connect the upper and lower platforms, and contribute to its impressive rigidity. The spindle, which virtually floats in space, can move at very high

Fig. 7-12-1 A vertical machining center with an automatic tool changer at the upper left. (*Cincinnati Machine, a UNOVA Co.*)

Fig. 7-12-3 On a universal machining center, the spindle can be programmed for a horizontal or vertical position whenever required. (*Deckel Maho, Inc.*)

Fig. 7-12-4 The VARIAX has a lower and upper platform and its spindle seems to be floating in space. (*Giddings & Lewis, Inc.*)

CNC ADVANTAGES

CNC machines require little operator intervention, and once the machine has been set up, it will machine without stopping until the end of the program is reached. Some of the other advantages that CNC gives a manufacturing shop are greater machine uptime, increased productivity, maximum part accuracy, reduced scrap, less inspection time, lower tooling costs, less inventory, and complex part production, Fig. 7-12-5.

Vertical machining centers are available in a variety of types and sizes but only the standard industrial and the Bench-Top teaching models will be covered in this book. For teaching purposes, the CNC Bench-Top machines will be used because they use the same basic programming features as industrial machines and they are relatively inexpensive and students feel at ease on a smaller machine, Fig. 7-12-6. References will be made to industrial machines throughout this unit because bench-top machines are so similar, especially those with Fanuc compatible controls. Programming codes vary slightly with different manufacturers, so it is always wise to consult the

speeds. The VARIAX has the potential to change and revolutionize CNC machine tools as they are known today.

The main advantages of the VARIAX are:

1. The machine can move and position the spindle in almost any direction giving it 6-axis contouring capabilities.
2. The rigidity of the triangulated crossed-leg structure is five times greater than that of a traditional machining center.
3. The machine can contour five to ten times faster than a conventional machining center.
4. The machine is two to ten times more accurate than a traditional machine tool.
5. The loads and movements are shared with all the actuating components, reducing the stress and wear in any one axis and providing **five times the** acceleration rate; for example, contouring can be performed at 2600 in/min.

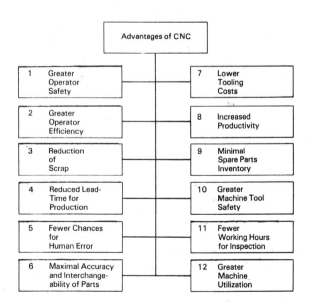

Fig. 7-12-5 The advantages of CNC have helped industry to improve productivity while also producing more accurate parts. (*Kelmar Associates*)

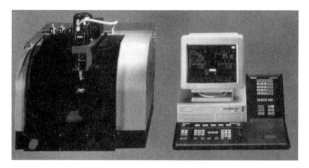

Fig. 7-12-6 The Bench-Top machining center is ideal for teaching CNC basics. *(Denford, Inc.)*

programming manual for each specific machine to avoid crashes or scrap work.

MACHINING CENTER PARTS

The main operating parts of CNC machining centers are the bed, column, saddle, table, servo motors, ball screws, spindle, tool changer, and the machine control unit (MCU), Fig. 7-12-7.

- The **bed**, usually made of high-quality cast iron, provides a rigid machine capable of performing heavy-duty machining and maintaining high precision. Hardened and ground bed ways provide rigid support for all linear axes movements.
- The **column**, mounted to the saddle, is designed to prevent distortion and deflection during machining. The column provides the machining center with the Z-axis vertical movement.
- The **saddle**, mounted on the hardened and ground ways, provides the machining center with X axis longitudinal linear movements.
- The **table**, mounted on the bed, provides the machining center with the Y axis cross-linear movement.

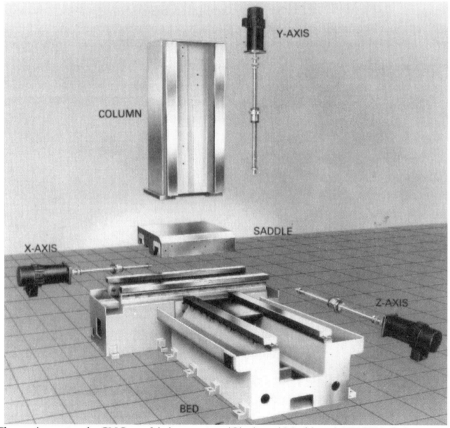

Fig. 7-12-7   The main parts of a CNC machining center. *(Cincinnati Machine, A UNOVA Co.)*

- The **servo system** consists of servo drive motors, ball screws, and position feedback encoders to provide fast, accurate movement and positioning of the XYZ axes slides.
- The **spindle**, programmable in one r/min increments, can have speed ranges from 20 to 6000 r/min. or higher. The spindle can be a fixed position (horizontal) type, or a tilting/contouring spindle, Fig 7-12-8, to provide an additional A axis movement.
- The **vertical tool changer**, Fig. 7-12-9, can store a number of preset tools that can be called for use by a command in the CNC part program. Tool changers are usually bi-directional and take the shortest travel distance to randomly access a tool. Modern tool change time is usually only 3 to 5 s, which improves machine uptime.

- The **machine control unit** (MCU) allows the operator to perform operations such as programming, machining, diagnostics, tool and machine monitoring. MCUs vary according to manufacturers' specifications and modern MCUs are more reliable and make the machining operations less dependent on the operator.

CARTESIAN COORDINATE SYSTEM

Standard Cartesian or rectangular coordinate system for CNC purposes may be likened to a grid arrangement of several square blocks of a community that may be used as reference points for giving directions. Each block acts as a unit of measurement in locating one point in the community in relation to another point. Assume that someone needs instructions to get from the pres-

Fig. 7-12-8 A CNC machining center with a tilting/contouring spindle provides the machine with an A axis. *(Cincinnati Machine, A UNOVA Co.)*

Fig. 7-12-9 The vertical tool changer can hold a series of tools for a job, each of which can be recalled as required by a call statement in the program. (*Fadal Engineering Co. Inc.*)

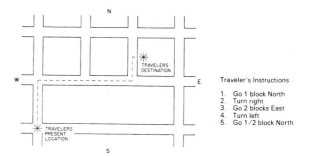

Fig. 7-12-10 A city street map containing a number of square blocks is a form of the coordinate system. (*Superior Electric Co.*)

ent location to their destination. By using the grid in Fig. 7-12-10 as a small sectional map, exact directions can be given from the present point to the destination point as listed to the right of the map. From this example, it can be seen that rectangular coordinates are often used in everyday living by expressing directions in two-dimensional coordinates.

The **Cartesian coordinates system** works on a grid system, similar to graph paper, where reference lines run at 90° to each other. If the center of the graph is considered as the X and Y zero or origin point, then all lines run parallel to either the X (horizontal) axis or Y (vertical) axis.

- If a point is plotted to the right of the X zero, the dimension would be a plus move or X+; to the left, it would be a minus move or X–.
- If a point is plotted above the Y zero, the dimension would be Y+; any point below Y zero would be Y–.

Plus dimensions do not have to be indicated on a CNC program with a plus (+) sign, they are assumed. The **minus (–)** sign must be included, otherwise the CNC control assumes that it is a plus move and locates the point above the X axis.

In Fig. 7-12-11, the graph paper has been divided into four quadrants, with the X and Y axes (origin point) in the center. The location of the three points, A, B, and C, is as follows:

- Point A is **X2** (two squares to the right of the Y axis) and **Y2** (two squares above the X axis).
- Point B is **X1** and **Y-2**
- Point C is **X-3** and **Y-3**

Machine Axes

Vertical machining centers, with their spindle in a vertical direction, fit the coordinate system very well because they generally have three primary axes of motion, the X (longitudinal), Y (cross), and Z (vertical), Fig. 7-12-12.

- The X motion on a CNC vertical machining center is the longitudinal movement of the table **RIGHT** or **LEFT**.
- The Y motion is the table cross-movement, **TOWARDS** or **AWAY** from the column.
- The Z motion is the vertical movement (**UP** or **DOWN**) of the spindle or knee.

Horizontal machining centers, Fig 7-12-13, which have their spindle in a horizontal direction, operate on three axes:

- The X axis controls the movement of the table, **RIGHT** or **LEFT**.
- The Y axis controls the vertical movement of the spindle, **UP** or **DOWN**.
- The Z axis controls the movement of the spindle, **IN** or **OUT**.

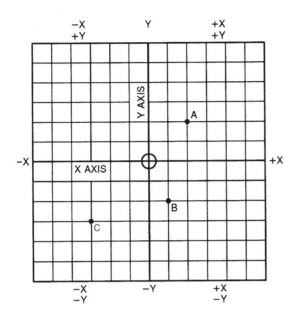

Fig. 7-12-11 The four quadrants formed when the X and Y axes cross allow points to be accurately located from the XY zero. *(Allen Bradley Co.)*

SUMMARY

• Machining centers evolved from the need to perform a variety of operations and machining sequences on a workpiece in one machine setup.

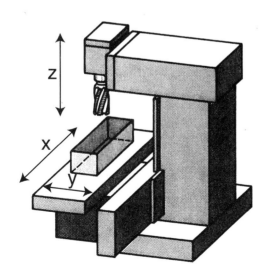

Fig. 7-12-12 The main axes of a vertical machining center. *(Deckel Maho, Inc.)*

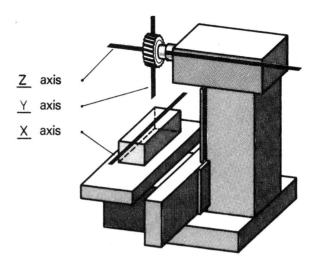

Fig. 7-12-13 The main axes of a horizontal machining center. *(Deckel Maho, Inc.)*

• The vertical, horizontal, universal, and Variax machining centers are widely used by industry. The vertical machining center is the most common.

• The advanced technological design of the Variax machine consists of six legs, upper and lower platform and a spindle that virtually floats in space.

• The main advantages of CNC machine tools are increased productivity, reduced scrap work, greater machine uptime, lower tooling and labor costs, and high-accuracy parts.

• The tool changer on machining centers makes it possible to include tool changes in the CNC program so that the correct tool is selected as required for the machining operation.

• The coordinate system used on machine tools makes it possible to locate any point within the grid system to a high degree of accuracy.

• Plus signs in a CNC program do not have to be included with the dimension, they are automatically assumed. Minus signs are included, otherwise the control will assume it to be a plus (+) dimension.

• The Z axis on a vertical machining center controls the spindle movement UP or DOWN. On a horizontal machining center, the Z axis moves the spindle IN or OUT.

KNOWLEDGE REVIEW

1. Why were machining centers developed?

Types of Machining Centers

2. Name four types of machining centers used in industry.
3. What type of machining center can change from a vertical to a horizontal type anytime during a machining program?
4. List four important advantages of the Variax.

CNC Advantages

5. Name five advantages of CNC machines.

Machining Center Parts

6. What purpose do the following parts serve?
    (a) column
    (b) saddle
    (c) table
    (d) tool changer

Cartesian Coordinate System

7. Why must negative (-) moves be indicated on a CNC program dimension?
8. Where is the XY zero or origin point located in the coordinate system?
9. Identify the following in relation to the X and Y axes:
    (a) Two squares to the left of the Y axis and three squares above the X axis.
    (b) Four squares to the right of the Y axis and four squares below the X axis.
    (c) Three squares to the left of the Y axis and four squares above the X axis.

Unit 13

CNC Bench-Top Mill

CNC bench-top teaching machines are widely used in high schools, community colleges, and industrial training programs to teach the basic fundamentals of CNC programming. The modern bench-top machines are usually equipped with Fanuc-compatible control units that have been universally accepted as the industry standard, Fig 7-13-1. Instructors and students have found these machines ideal for teaching and learning purposes because of their small size and simplicity of operation. They are relatively inexpensive, programmed in the same way by using the same G and M codes as industrial-size machines, but use smaller workpieces and take lighter cuts.

OBJECTIVES

After completing this unit, you should be able to:

1. Be familiar with the common **G** and **M** programming codes used on CNC mills.
2. Understand the main datum point symbols used on CNC bench-top milling machines.
3. Coordinate the setting of the cutting tool, cutter offsets, and the workpiece for accurate machining.
4. Program a basic part that uses linear and circular interpolation.

KEY TERMS

climb milling	function codes
conventional milling	machine axes
cutter diameter compensation	preparatory codes
datum points	work coordinates

CNC MACHINE AXES

Machining centers have probably made the greatest impact in CNC machining because of their ability to perform a wide variety of operations on a workpiece in one setup. The CNC milling machine is one of the most versatile machine tools used in industry, Fig. 7-13-2. Operations such as flat and straight milling, contouring, drilling, boring, and reaming are a few of the operations that can be performed on a CNC mill. Most bench-top teaching size CNC mills can be programmed on three axes:

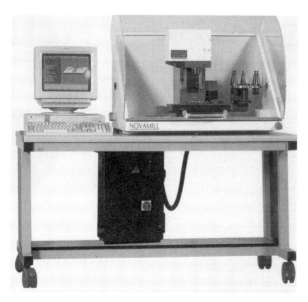

Fig. 7-13-1 Bench-top CNC machines are ideal for teaching the basic fundamentals of CNC programming. *(Denford Inc.)*

- The X axis controls the table movement left and right.
- The Y axis controls the table movement toward or away from the column.
- The Z axis controls the vertical (up or down) movement of the knee or spindle.

THREE AXIS MACHINES

The three axis CNC bench-top milling machine, equipped with the Fanuc controller, can use all important G and M programming codes, milling cycles, subroutines, etc. These machines can be programmed in inch or metric dimensions using either absolute or incremental programming. Some models can be set for both vertical and horizontal machining. They may also be equipped with a graphics display to allow the operator to check the accuracy of the program visually on the computer screen before actually cutting the part, Fig. 7-13-3.

MILLING CUTTER USE

To get the best tool life and productivity out of a milling cutter, it is important that the cutter be used properly. Most milling tool problems are

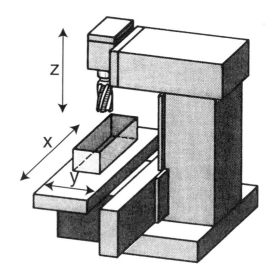

Fig. 7-13-2 The main axes of a CNC vertical machining center. *(Denford Inc.)*

caused by excessive heat, abrasion, edges chipping, clogging, built-up edges, cratering, and work hardening of the workpiece. Before using any cutter, it is important to decide whether climb or conventional milling should be used to perform the operation.

Climb milling takes place when the cutter rotation and the table (work) feed are in the same direction, Fig. 7-13-4 A. When climb milling, it is important that the work is held securely, and the machine table should be equipped with a backlash eliminator. On conventional machines, the table-slide gibs should be tightened enough to prevent the cutter from being drawn or climbing into the work.

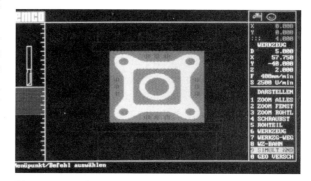

Fig. 7-13-3 The 3D simulation on a computer screen of the tool path and machining operations for milling a part. *(Emco Maier Corp.)*

Conventional milling takes place when the cutter rotation and the table (work) feed are going in opposite directions, Fig. 7-13-4B. The tooth enters the cut at zero chip thickness in an upward direction and progressively gets thicker. Conventional milling is used on workpieces where minimum shock is desirable when the cutter enters the work, or when the conventional machine table is **not fitted with a backlash eliminator.** The forces in conventional milling try to lift the workpiece; therefore it is important that the work be held securely.

Basic Cutter Practice

The following are a few of the points that should be observed when using milling tools:

1. Tighten locks screws on the cutter shank if the adaptor is so equipped, to provide a solid drive.
2. When mounting cutters, be sure that the adaptors, arbors, collets, and cutters are clean and free of dirt and burrs that could cause misalignment.
3. Check for cutter run-out and correct the problem. Run-out could cause undue wear on a small section of the cutter.
4. Remove as much backlash as possible from conventional table feed screws by

snugging up the table-slide gibs to reduce play and dimensional errors while machining.
5. Use the low range of the recommended speeds and feeds on new applications. These can be increased gradually to increase productivity when suitable.
6. **Never machine with a dull cutter.** It is wise to determine just how many pieces a cutter might be expected to produce before it should be resharpened.
7. Use coolant wherever possible to reduce the machining friction and heat that will shorten the life of a cutting tool.

MACHINE CODES

There are two main types of machine codes used in CNC programming, the preparatory codes or functions and the miscellaneous codes or functions.

Preparatory Functions

The **preparatory functions or cycle codes,** commonly called **G codes**, refer to a mode of operation of the machine tool or CNC system. It generally refers to some action on the **X, Y,** and/or **Z** axes.

- In CNC programming, the word address letter G refers to a preparatory function and is followed by a two-digit number; e.g., a G00 function would be point-to-point positioning at a rapid rate of about 700 to 2000 in./min (18 to 50 m/min).
- Common preparatory functions include operations such as point-to-point positioning, linear interpolation, circular interpolation, absolute or incremental programming, inch or metric programming, and fixed (canned) cycles. Each of these is designated by a G code, which the central processing unit (CPU) and the machine tool can recognize and act upon accordingly.

The most commonly used preparatory func-

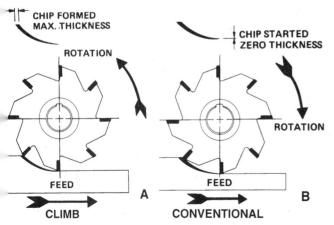

**Fig. 7-13-4 The two common methods of milling. (A)
Climb. (B) Conventional.** *(Niagara Cutter, Inc.)*

tions in Fig. 7-13-5 have been supplied by the EIA and are listed in their standard EIA-274-D.

Miscellaneous Functions

Miscellaneous CNC functions, commonly called **M codes**, perform a variety of auxiliary commands, such as stopping the program, starting or stopping the spindle or feed, tool changes, coolant flow, etc., that control the machine tool. They are generally multi-character **ON/OFF** codes that select a function controlling the machine tool. Miscellaneous functions are used at the beginning or end of a cycle and are identified by the letter address **M** followed by a two or three digit number.

Miscellaneous codes such as M00, M01, M02, or M06 are usually effective only in the specific block in which they are programmed. Most other miscellaneous codes do not have to be repeated in succeeding blocks.

Modal Codes

Both preparatory and miscellaneous functions or codes are generally classified as either *modal* or *nonmodal*. When *modal miscellaneous functions* such as M03 (spindle ON CW) and preparatory functions such as G81 (drill cycle) are programmed, they stay in effect in succeeding blocks until they are replaced by another function code. The modal function or code is changed or canceled as soon as a new preparatory function code is programmed.

All *nonmodal functions* such as M00, M01, M02, M06, etc., are valid or operational only in the block programmed. If they are needed in successive blocks, they must be programmed again.

The miscellaneous function codes in Fig. 7-13-6 have been supplied by the EIA and are listed in their standard ElA-274–D.

Group	G code	Function
01	G00	Rapid positioning
01	G01	Linear interpolation
01	G02	Circular interpolation clockwise (CW)
01	G03	Circular interpolation counterclockwise (CCW)
00	G04	Dwell
02	G17	*XY* plane selection
02	G18	*ZX* plane selection
02	G19	*YZ* plane selection
06	G20	Inch input (in.)
06	G21	Metric input (mm)
00	G27	Reference point return check
00	G28	Return to reference point
00	G29	Return from reference point
07	G40	Cutter compensation cancel
07	G41	Cutter compensation left
07	G42	Cutter compensation right
08	G43	Tool length compensation in positive (+) direction
08	G44	Tool length compensation in minus (−) direction
08	G49	Tool length compensation cancel
09	G80	Canned cycle cancel
09	G81	Drill cycle, spot boring
09	G82	Drilling cycle, counterboring
09	G83	Peck drilling cycle
09	G84	Tapping cycle
09	G85	Boring cycle #1
03	G90	Absolute programming
03	G91	Incremental programming
00	G92	Setting of program zero point
05	G94	Feed per minute

Fig. 7-13-5 Commonly used milling preparatory function codes according to the EIA Standard – 274-D.

Code	Function
M00	Program stop
M01	Optional stop
M02	End of program (no rewind
M03	Spindle start (forward CW)
M04	Spindle start (reverse CCW)
M05	Spindle stop
M06	Tool change
M07	Mist coolant on
M08	Flood coolant on
M09	Coolant off
M19	Spindle orientation
M30	End of program (return to top of memory)
M48	Override cancel release
M49	Override cancel
M98	Transfer to subprogram
M99	Transfer to main program (subprogram end)

Fig. 7-13-6 The most common EIA codes that control miscellaneous machine control functions.

COORDINATES—MACHINE AND WORK

All CNC machine tools have a default coordinate system (based on a home or zero position) that is a fixed position built into the machine by the manufacturer. The zero position is usually located at the tool-change position and could be at the upper right hand corner or the upper left corner depending on the manufacturer, Fig. 7-13-7.

The zero position is used as a reference point when locating work-holding devices and workpieces. Most MCUs have two position registers, the machine position and the command position. The machine position register shows the position of the machine table from the reference zero or home position at all times. The command position register shows the table axis position in relation to the programmer's or operator's **XY** zero or reference point.

When a part is being programmed, the machine's reference or zero position and the part zero are rarely the same. The programmer can choose any suitable location on the part, work-

Stored zero shifts (G54...G59)
Programmed zero shift (G92)

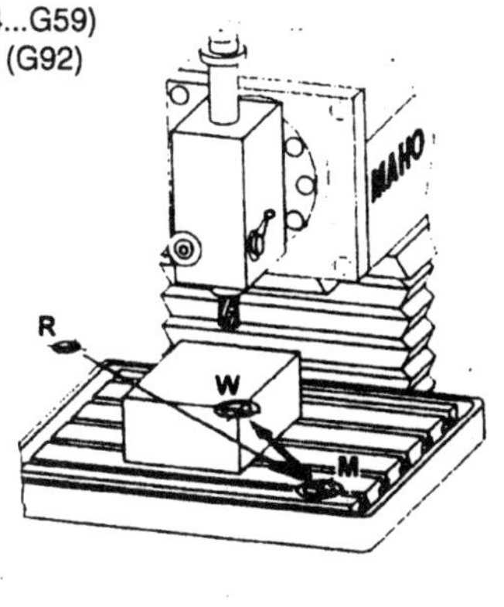

⊕ R = Reference point (maximum travel of machine)
⊕ M = Machine zero point (X0,Y0,Z0) of machine coordinate system.
⊕ W = Part zero point workpiece coordinate system.

Under G54 ... G59 the actual machine coordinates of part zero are stored in the stored zero offsets memory and activated in the part program.

Under G92 the actual machine coordinates are inserted and used on the G92 line of the part program.

Fig. 7-13-7 **The relationship between the part zero and the machine system of coordinates.** *(Deckel Maho, Inc.)*

holding device, or machine table that is best suited as the program zero.

Part Zero Point

For convenience in programming, the machine zero point can be set by three methods: by the operator, by a programmed absolute zero shift, or by work coordinates to suit the workholding fixture or the part to be machined.

Manual Setting - The operator can use the controls on the machine control unit (MCU) to locate the spindle over the desired part zero and then set the **X** and **Y** coordinate registers on the console to zero.

Absolute Zero Shift - The absolute zero shift is a method of changing the coordinate system by a G92 command in the CNC program. The programmer first sends the machine spindle to home zero position by a G28 command in the program. This step is followed by another command (G92 for absolute zero shift) to tell the MCU how far from the home zero location the coordinate system origin is to be positioned, Fig. 7-13-7.

The sample commands may be as follows:

N010 G91 G28 X0 Y0 Z0 (sends spindle to home zero position)
N020 G92 X3.000 Y4.000 Z5.000 (the position the machine will use as part zero)

If two or more fixtures or parts are to be located on a machine table, the programmer can use a different zero location for each part or fixture. This setting is done by inserting a G91 G28 X0 Y0 Z0 command to return to home zero, followed by a G92 command with new coordinates for each part or fixture zero location required. For safety reasons, often the **Z** axis machine zero return is performed before the **X** and **X** axes.

Work Settings and Offsets

All CNC machine tools require some form of work setting, tool setting, and offsets (compensation) to place the cutter and work in the proper relationship. Compensation allows the programmer to make adjustments for unexpected tooling and setup conditions.

Work Coordinates

In absolute positioning, work coordinates are generally set on one corner of a part and all programming is taken from this position. In Fig. 7-13-8, the part zero is used for all positioning for hole locations 1, 2, and 3.

In incremental positioning, the work coordinates change because each location is the zero point for the move to the next location, Fig. 7-13-9.

On some parts, it may be desirable to change from absolute to incremental, or vice versa, at certain points in the job. By inserting the G90 (absolute) or the G91 (incremental) command into the program at any point, absolute mode can be the changed to incremental mode or vice versa.

Work Plane or Gage Height

In a fixed cycle, the word address letter **R** refers to either the work surface or the rapid-feed distance (level) from which the feed rate is applied. The **R** level is set at a specific height (distance) above the part surface in the program. This surface setting is referred to as **R** level, or the clearance dimension. All absolute programmed depths for cutting tools are taken from the **Z0** of the part.

In a fixed cycle, the **R** work plane (**R** level) is generally established at .100 in. above the cutting surface of the workpiece, Fig. 7-13-10. This surface is also known as gage height. When cutting tools are set up, the operator generally touches the tool tip on top of the highest surface of the workpiece, and all tools (regardless of length) are set to this height. Once the tools have been set, no further adjustment is necessary since most MCUs automatically add the measured tool length to all future depth dimensions.

TOOL SETTINGS AND OFFSETS

All types of CNC machine tools require some form of tool setting and offsets (compensation) to allow the programmer or operator to meet unexpected problems relating to tooling. It often is impossible to predict the exact tooling problems that may come up on a job. Tool compensation allows changes to be made to the program to overcome any tooling problems that arise.

Setting the Zero Point and Tool Axes

Before any part is machined it is very important to set the cutting tool to the zero locations of the **X**, **Y**, and **Z** axes. Every machine tool has a specific home or zero reference point that is built into it by the manufacturer and cannot be changed. In

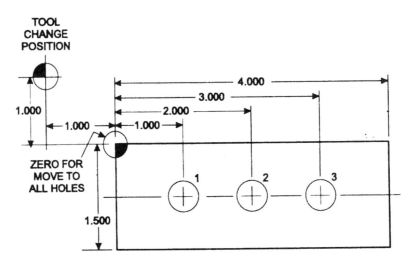

Fig. 7-13-8   In absolute programming, all dimensions are taken from the XY zero at the top left-hand corner of the part. *(Kelmar Associates)*

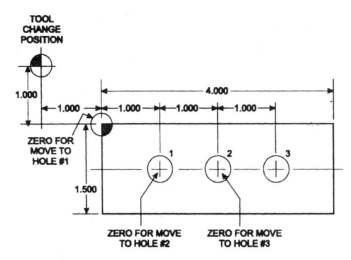

Fig. 7-13-9 In incremental programming, all dimensions are taken from the previous point. *(Kelmar Associates)*

Fig. 7-13-11, the **XY** zero (home) position of the machine is at the top right-hand corner of the table's working range. However, the programmer can temporarily change this position to any point that best suits the part, operation, or tool-change position. The programmed reference point will not usually be the same as the machine reference point.

Setting X and Y Axes to Zero

After a part has been fastened to the table and aligned, the center of the machine spindle should be aligned with the finished **X** and **Y** edges of the part.

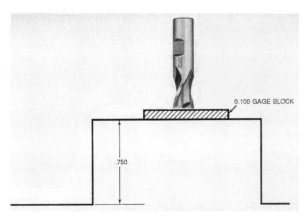

Fig. 7-13-10 Using a .100 in. gage block to set the gage height or R level on the work surface.
(Kelmar Associates)

Using an edge finder with a .200 in. diameter for location purposes, follow the procedure below:

To Set the X Axis to Zero

1. With the edge finder held in the machine spindle, lower the spindle until the diameter of the edge finder is close to the left end of the part along the **X** axis, Fig. 7-13-12.
2. With a thin piece of paper between the edge finder and the end of the part jog the table along the **X** axis until it just contacts the paper.
3. Raise the spindle until the edge finder clears the top of the workpiece.
4. Jog the table the thickness of the paper in the direction of the part.
5. Move the table one-half the diameter of the edge finder, or .100 in., along the **X** axis.
6. Set the **X** axis register to zero.

To Set the Y Axis to Zero

The procedure used for setting the center of the machine spindle on the **X** axis should be used on the top edge to set the **Y** axis, Fig. 7-13-13. Once this is done, be sure to set the **Y** register to zero. The zero shift on the machine control unit (MCU) allows the **XY** zero location to be easily shifted to the one chosen by the programmer.

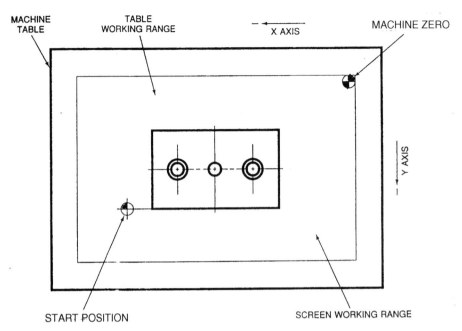

Fig. 7-13-11 The machine XY zero (upper right corner) is a fixed position, and the command or part zero can be moved anywhere within the working range of the machine table. *(Kelmar Associates)*

Setting the Z Axis to Zero

Use the following procedure for setting the end of the milling cutter to the workpiece surface in the **Z** axis:

1. Move the table until the cutter is over the highest surface of the part (**Z0**, Fig. 7-13-14.
2. Place a thin piece of paper between the surface of the work and the end of the cutter.
3. Jog the slide in the **Z** axis until a slight drag is felt on the paper.

4. Move the table in the **X** or **Y** direction so that the cutter clears the workpiece.
5. Lower the spindle the thickness of the paper, approximately .002 in.
6. Set the **Z** axis register to zero.

WORK SETTINGS AND OFFSETS

All CNC machine tools require some form of tool setting, work setting, and radius compensation.

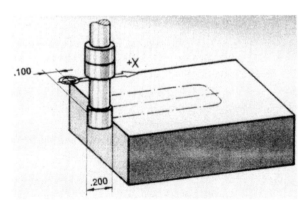

Fig. 7-13-12 Setting the spindle center to zero on the left X axis of the part with an edge finder.
(Deckel Maho, Inc.)

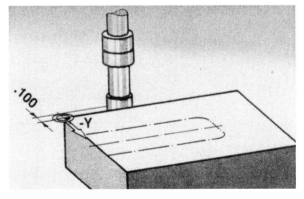

Fig. 7-13-13 Setting the spindle center to zero on the top Y axis of the part with an edge finder.
(Deckel Maho, Inc.)

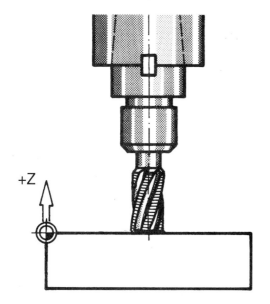

Fig. 7-13-14 Setting the bottom of the end mill to the work surface in the Z axis. *(Deckel Maho, Inc.)*

Compensation allows the operator or programmer to make adjustments for tooling and setup conditions that may arise during setup or machining operations. It is difficult to foresee all the conditions that could arise and compensation is designed to allow the operator to handle unexpected problems.

Cutting Tools

For every part, the programmer must plan each operation required to machine the part and also the cutting tools required for each operation. The tool list must include the type of each cutting tool, its diameter, and length. Since a wide variety of cutting tools of various lengths and diameters are generally used for machining a part, it is important to be able to make allowances for various diameters and lengths to ensure that the part will be machined correctly. This compensation involves working with offsets so that the machine control unit knows exactly how to adjust for differences in tool diameters and lengths.

CNC offsets can be compared to the memory on an electronic calculator. If the calculator has a memory, a constant value can be stored into each memory and used whenever required for a cal-culation. This keeps from having to enter the same number over and over again when it is required. Like the file locations in the memory of an electronic calculator, offsets in the CNC control are storage locations into which numerical values for each tool can be placed.

Tool Lengths

It is the responsibility of the operator to set a minimum tool length (tool extension), based on the depth to which each cutting tool must travel. Each tool is then programmed as though it had zero length, and the spindle/tool gage line and the tool point were the same. The amount of **Z** axis travel is programmed from the **Z0**, usually starting .100 in. above the work surface, at the programmed feed rate, Fig. 7-13-15.

The actual (measured) length of each tool is entered into the MCU through the computer, or manually by the operator.

Some MCUs are equipped with semi-automatic tool compensation to make allowances for differences in tool lengths. The tool-length compensation feature provides for as many as sixty-four or more tool length offsets. When the

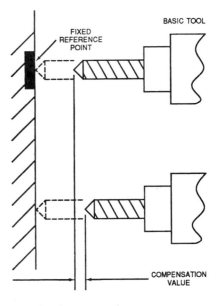

Fig. 7-13-15 Semi-automatic compensation is used to calculate tool lengths for all tools at setup time. *(Modern Machine Shop Magazine)*

Setting zero point in the tool axis

The length of the various tools to be used for each operation must be recorded and entered into the tool memory

In actual practice **two modes of operation** are used.

1. Machining with **one** tool

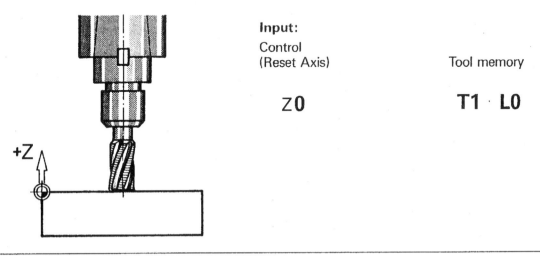

Input:
Control
(Reset Axis)

Z0

Tool memory

T1 · L0

2. Machining with **several** tools

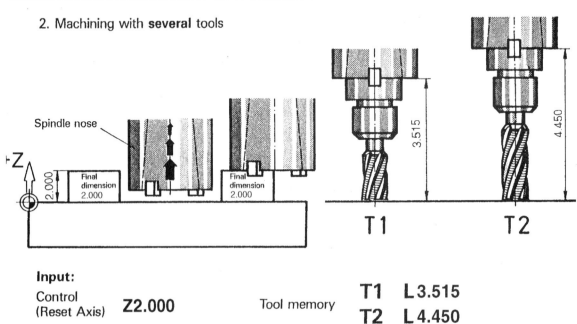

Input:
Control
(Reset Axis)   **Z2.000**   Tool memory

T1   L3.515
T2   L4.450

Fig. 7-13-16 Two methods of setting and recording variations in tool lengths into the MCU memory.
(Deckel Maho, Inc.)

machine is being set up, the operator mounts each tool in succession, starting with the longest, jogs it against a fixed reference stop, and presses the tool-offset button on the MCU. This procedure automatically records the tool's setting into memory. All cutting tools for a job are set against the same reference stop, and the difference between each tool's actual length and the basic tool length is entered into memory as a compensation value, Fig. 7-13-16. With this information, the control's computer changes all program steps to take into account the length and diameter of each tool.

Some CNC machines use preset tools that have been prepared in the toolroom. Their type, diameter, and length have all been recorded on the tool-assembly drawing and used when the program is being prepared. Each tool is assigned a specific number that is stored in the MCU, and can be recalled by a command in the CNC program any time the tool is required. For example, a T1 M06 code will place tool #1 into the spindle on a machine equipped with an automatic-tool changer (ATC). Newer CNC machines are capable of detecting tool-length compensation.

CUTTER RADIUS COMPENSATION

Cutter-radius compensation (CC) changes a milling cutter's programmed centerline path to compensate for a small difference in cutter diameter. On most modern MCUs, it is effective for most cuts made using either linear or circular interpolation in the XY axis, but does not affect the programmed Z axis moves. Compensation is usually in increments of .0001 in. up to 1.0000 in., and most modern controls have at least as many compensation registers available as there are tool pockets in the tool-storage magazine.

The advantages of the cutter-radius compensation feature are that it:

1. Allows use of cutters that have been sharpened to a smaller diameter.
2. Permits use of a smaller or larger tool in the machine's storage magazine.
3. Allows the tool to be backed away when

roughing cuts are required due to the presence of excessive material.
4. Permits compensation for unexpected tool or part deflection, if the deflection is constant throughout the programmed path.

The basic reference point of the machine tool is never at the cutting edge of a milling cutter, but at its center point. If a 1.000 in. end mill is used to machine the edges of a workpiece, the programmer has to keep a .500 in. offset from the work surface in order to cut the edges to the correct size, Fig. 7-13-17. The .500 in. offset represents the distance from the centerline of the cutter or machine spindle to the edge of the part. On older CNC machines the programmer must calculate an offset path, which is usually half the cutter diameter.

Modern MCUs, that have part surface programming, calculate centerline offsets automatically once the radius of the cutter for each operation is stored and called by the program. Many MCUs have operator-entry capabilities that can be used to compensate for differences in cutter diameters; therefore an oversize cutter, or one that has been sharpened, can be used as long as the compensation value for oversize or undersize cutters is entered.

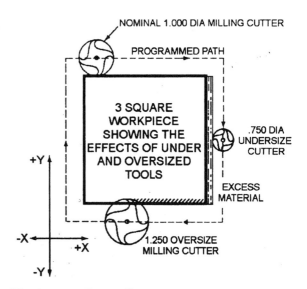

Fig. 7-13-17 Cutter diameter compensation must be used when machining with various sizes of cutters. (Kelmar Associates)

The principle of the tool radius offset

The cutter radius offset makes it possible to program the workpiece dimensions.

Principle:

Tool memory:

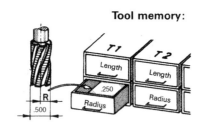

The workpiece dimensions are programmed and the control calculates the tool center path by means of the respective radius value R.

Input into the tool memory:

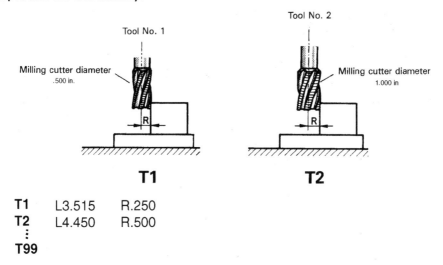

T1	L3.515	R.250
T2	L4.450	R.500
⋮		
T99		

Fig. 7-13-18 The principle of cutter-radius offset. *(Deckel Maho, Inc.)*

Programming Cutter-Radius Compensation

The use of cutter-radius compensation may vary with different controls. Each control generally has a set of rules that specify how cutter-radius compensation is used and canceled on that control. The basics of how a control is programmed are shown in the following example. Be sure to refer to the CNC control manufacturer's manual for the machine control unit being used.

Most controls use three G codes with cutter-radius compensation. **G41** is used to activate a cutter left condition (climb milling with a right hand cutter). **G42** is used to activate a cutter right condition (conventional milling with a right hand cutter). **G40** is used to cancel cutter-radius compensation. The principle of cutter-radius offset is illustrated in Fig. 7-13-18.

To know when to use G41 or G42 code, look in the direction the cutter is moving during machining.

- When the cutter is on the **left of the work**, use the **G41** code.
- When the cutter is on the **right of the work**, use **G42** code.

Figure 7-13-19 shows some examples that should help in deciding when to use the G41 or G42 code. Once cutter-radius compensation is properly activated, the cutter will be kept on the

G41/G42, G40

To enable the control to calculate the center path of the milling cutter from the program data and from the tool memory data, it is necessary to communicate to the control where the tool has to cut.

For this setting there are 3 G-functions:

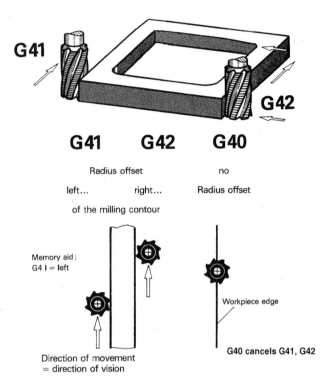

Fig. 7-13-19 Methods of programming cutters for radius compensation. *(Deckel Maho, Inc.)*

same side of all surfaces until the G40 command is used to cancel the compensation.

CNC PROGRAMMING HINTS - MILLING

 Machine reference point (maximum travel of machine).

 Machine **XY** zero point (could also be tool-change point).

 Part **XY** zero point (programming start point).

 Indicates the tool-change position. A G92 code will reset the axis register position coordinates to this position.

For a program to run on a machine, it must contain the following codes:

M03 To start the spindle revolving.

Sxxx The spindle speed code to set the r/min.

Fxx The feed rate code that will move the cutting tool or workpiece to the desired position at the desired rate.

ANGLES:

The **XY** coordinates of the start and end point of the angular surface plus a feed rate **(F)** are required.

Z CODES:

- A **Z positive** dimension raises the cutter above the work surface.
- A **Z negative** dimension feeds the cutter into the work surface.
- **Z.100** is the recommended retract distance above the work surface before a rapid move **(G00)** is made to another location.

ARC/CONTOUR REQUIREMENTS:

- The start point of the arc **(XY coordinates)**.
- The direction of cutter travel **(G02 or G03)**.
- The end point of the arc **(XY coordinates)**.
- The center point of the arc **(IJ coordinates)** or the **(arc radius R)**.

MILLING AND DRILLING PROGRAMMING EXERCISE

Program Notes: (Fig. 7-13-20)

- Program in the absolute mode starting at the tool-change position at the top left corner of the print.
- The material is aluminum (300 CS), feed rate 10 in/min.
- The cutting tool is a .250 in. diameter high-speed steel 2–flute end mill.
- Mill the 1.000 in. square slot.
- Drill the two .250 in. diameter holes, .250 in. deep.
- Mill the .250 in. wide angular slot, .125 in. deep.

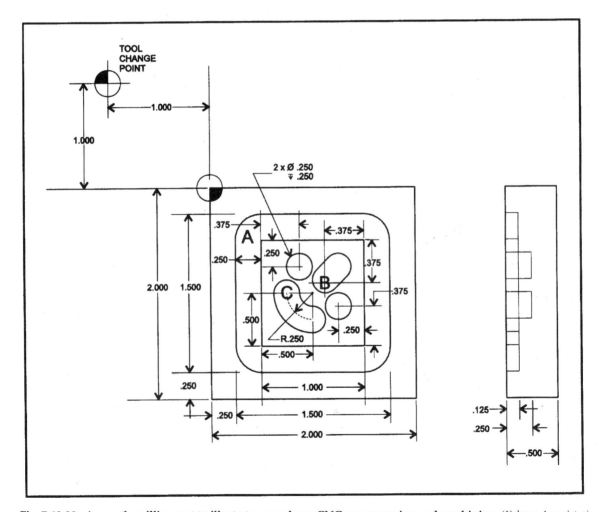

Fig. 7-13-20   A sample milling part to illustrate procedures CNC programming and machining. *(Kelmar Associates)*

- Mill the .250 in. wide circular groove, .125 in. deep.
- After the job is completed, return to the tool-change position.

Programming (tool is in the spindle)

%	(rewind stop code/parity check)
O2000	(program number)

(T01 = Tool 1 - .250 diameter, two-flute end mill)

N010 G92 G54 X-1.000 Y1.000 Z1.000

G92	programmed offset of reference point (tool-change position)
G54	work offset
X-1.000	tool set at 1.000 to the left of the part.
Y1.000	tool set at 1.000 above the top edge of the part.
Z1.000	the end of the cutter is 1.000 above the top surface of the part.

N020 G20 G90

G20	inch data input.
G90	absolute programming mode.

N030 M06 T01

M06	tool change command.
T01	tool no. 1 (.250 diameter, 2–flute end mill).

N040 S2000 M03

S2000	spindle speed set at 2000 r/min.
M03	spindle on clockwise.

N050 G00 X0 Y0 Z.100

G00	rapid traverse rate to X0 Y0 at the top left corner of the part.
Z.100	tool rapids down to within .100 of the work surface.

Machining the square groove

N060 X.375 Y-.375 (tool rapids to position A).

N070 G01 Z-.125 F10

G01	linear interpolation.
Z-.125	tool feeds .125 below the work surface.
F10	feed rate set at 10 in./min.

N080 X1.625 Y-.375

X1.625	top groove cut to the right hand end.

Y-.375	measurement did not change because it was set in block N060.

N090 Y-1.625

Y-1.625	right hand side of the groove cut.

N100 X.375

X.375	bottom groove cut to the left side.

N110 Y-.375

Y-.375	left-hand side of groove cut; this completes the groove.

N120 G00 Z.100

G00	rapid traverse mode.
Z.100	tool rapids to .100 above work surface.

Hole Drilling

N130 G00 X.875 Y-.750

tool rapids to the top left hole location.

N140 G01 Z-.250 F10

tool feeds .250 into work at 10 in./min. to drill the first hole.

N150 G00 Z.100

tool rapids out of hole to .100 above work surface.

N160 X1.250 Y-1.125

tool rapids to second hole location.

N170 G01 Z-.250 F10

tool feeds .250 into work at 10 in./min. to drill the second hole.

N180 G00 Z.100

tool rapids out of hole to .100 above work surface.

Machining the Angular Slot

N190 X1.125 Y-.875 (location B)

tool rapids to the start of the angular slot.

N200 G01 Z-.125 F10

G01	linear interpolation.
Z-.125	tool feeds to .125 below the work surface.
F10	feed rate set at 10 in./min.

N210   X 1.250 Y-.750

> angular slot cut to top right corner.

N220   G00 Z.100

> tool rapids to .100 above work surface.

Machining the Circular Groove

N230   X.750 Y-1.000 (location C)

> tool rapids to start of circular groove.

N240   G01 Z-.125 F10

> tool feeds to .125 below the work surface.

N250   **G03 X1.000 Y-1.250 R.250**

> G03   circular interpolation counterclockwise.
> X&Y   location of end of circular groove.
> R.250   radius of arc is .250.

N260   **G00 Z.100**

> tool rapids to .100 above work surface.

N270   X-1.000 Y1.000 Z1.000

> tool rapids back to tool-change position.

N280   M05   **spindle turned off.**
N290   M30   end of program

SUMMARY

- The three main axes of CNC Machining Centers are the **X** axis that controls the table movement left and right, the **Y** axis that controls the table towards or away from the column, and the **Z** axis that controls the spindle movement up or down.
- Climb milling is when the cutter rotation and the table feed are in the same direction; conventional milling is when cutter rotation and the table feed are in opposite directions.
- Preparatory functions (G codes) control the operation of the machine tool or CNC system; miscellaneous functions (M codes) control a variety of auxiliary commands.
- Modal codes stay in effect until they are replaced by another function code; nonmodal codes are operational only in the block programmed.
- The main coordinate system locations are the machine zero point, tool-change point, and the work zero point.
- All CNC machine tools have some form of tool setting and offsets to allow the programmer to compensate for variations in tool diameter and length.
- Before any machining is performed, it is necessary to set the **X** and **Y** axes to the work zero, and the **Z** axis to the highest surface of the workpiece.
- Cutter radius compensation changes a milling cutter's programmed centerline path to compensate for difference in the tool diameter.
- G41 code activates a cutter left location, G42 activates a cutter right location, and G40 cancels the cutter-radius compensation.

KNOWLEDGE REVIEW

CNC Machine Axes

1. On CNC mills, what do the following axes control?
   (a) **X**   (b) **Y**   (c) **Z**

Milling Cutter Use

2. Name seven causes of the most common milling tool problems.
3. Define: (a) climb milling   (b) conventional milling

4. What could cause misalignment when mounting adaptors, collets, and cutters?

Machine Codes

5. To what does a preparatory or G code refer?
6. Name the preparatory codes used for the following:
   (a) linear interpolation
   (b) circular interpolation clockwise

(c) inch input
(d) absolute programming
7. What is the purpose of miscellaneous or M codes?
8. Name the preparatory codes used for the following:
(a) spindle start clockwise
(b) spindle stop
(c) tool change
(d) end program, return to top

Coordinates – Machine and Work

9. Identify the following symbols:

(a)   (b)   (c)

10. Name three methods that can be used to set the machine zero point.
11. Define: (a) absolute positioning, (b) incremental positioning.
12. To what distance is the R level usually set above the part surface?

Tool Settings and Offsets

13. Where is the **XY** zero (home) position of a machine tool generally located?
14. To what finished edges should the center of a machine spindle be aligned?
15. List the steps required to set the Z axis to zero.

Work Settings and Offsets

16. What information about a cutting tool should be included on a tool list?
17. What is the purpose of semiautomatic tool compensation?
18. State the advantages of using preset tools on CNC machines.

Cutter Radius Compensation

19. What is the purpose of cutter-radius compensation?
20. What is the function of MCUs containing part surface programming?
21. What cutter-radius compensation codes are used for
(a) cutter right location?
(b) cancel cutter-radius compensation?
(c) cutter left location?

CNC Programming Hints – Milling

22. List the three codes that are necessary for a CNC program to run on a machine.
23. What four pieces of information are necessary in a CNC program in order to machine a radius or contour?

Unit 14

CNC Machining Center Programming

The first version of the CNC machining center, a three axis, vertical spindle Hydrotel, was introduced in the early 1950's. This machine was capable of machining parts using simultaneous three-axis cutting tool movement. This design was a major advance over conventional milling machines. However the machine operator still had to monitor the cutting tool performance, change dull or broken tools manually, and set the cutting speeds and feeds. These problems were a concern to the operator and the machine tool manufacturer who continually worked to improve the design and operation of the machining center.

CNC machining centers have become one of the most important machine tools used in industry. They vary in design from simple knee type milling machines to a five-axes profiler. They vary in size and features, but they all have one thing in common, their primary axes are the **X** and **Y** axes.

OBJECTIVES

After completing this unit, you should be able to:

1. Know why tool length compensation is needed when setting up a machining center.
2. Know the uses of cutter-radius compensation.
3. Understand the use of cutter-radius offsets and where they are stored in the machine control unit.
4. Write a CNC program for a machining center.

KEY TERMS

cutter-radius compensation

cutter-radius offset

simultaneous cutting

tool length compensation

MACHINE TYPES

The three most common types of machining centers are the:

• CNC Vertical Machining Center (VMC)

Fig. 7-14-1 CNC machining centers are capable of performing a wide variety of operations.
(Cincinnati Machine, A UNOVA Co.)

• CNC Horizontal Machining Center (HMC)
• CNC Horizontal Boring Mill

The CNC machining center can be divided into three main categories:

• By the axis orientation – vertical or horizontal.
• By whether the machine has a tool changer or no tool changer.
• By the number of axes.

Machining centers where the axial spindle motion is designed to move up and down are classified as vertical machines. On horizontal machines the axial spindle motion is designed to move horizontally. These simple definitions basically describe the difference between vertical and horizontal machining centers, but do not reflect the current state of the art machine tools. The machine tool industry is constantly striving to produce new and more powerful machine tools and designs for manufacturers, (for example, the Variax machine) Fig. 7-12-14.

MACHINE ACCESSORIES

Modern CNC machining centers are designed to perform a wide variety of machining operations along with traditional milling. They are also capable of operations such as drilling, reaming, boring, tapping, profiling, and thread cutting. These machines may be equipped with a tool magazine (also known as a carousel), a fully automatic tool changer (ATC), and a powerful computerized control unit (CNC). Some designs may have an automatic pallet changer (APC), adaptive control, robot interface, probing system, and high-speed machining features. Machine tools with these capabilities cannot be classified as simple CNC milling machines. Milling machines having at least some of the advanced features have become new machine tools - the machining centers.

MACHINE AXES

Standard CNC machining centers have three axes, the **X**, **Y**, and **Z** axes. More advanced machining centers may have a fourth axis (A-axis for vertical machines, or B-axis for horizontal machines), usually consisting of an indexing or a rotary axis, Fig. 7-14-2. Some machines may have five or more axes for doing specialized work that requires simultaneous cutting motion for machining complex shapes and cavities.

Vertical Machining Center Axes

Many types of machining centers are identified by the number of axes in which each type is capable of operating. A brief description of the most common types follows:

- Two and one half axis machines where all three axes (**X**, **Y**, and **Z**) cannot be operated simultaneously. These were the early types of CNC machining centers.
- Three axis machines where all three axes can be operated at the same time. This is the most

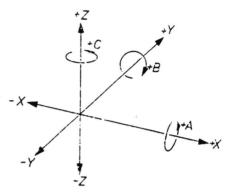

Z-Axis Vertical

Fig. 7-14-2   Linear and rotary axes used on machining centers. *(Allen Bradley)*

common type used for performing standard operations.
- Four axis simultaneous machines are the same as three axis machines with the addition of a rotary axis (A), usually in the form of a rotary table or indexing head.
- Five or more axis machines are the same as four axis machines with additional axes for machining complex shapes or contours.

The programming methods do not vary much for either machine type, except for special accessories and options. Some of the major differences are:

- The orientation of the machine axes.
- Additional axes for indexing or full rotary motion.
- The type of work suitable for individual models.

MACHINE ZERO POINTS

The machine zero point is a fixed location that is set by the machine tool builder and cannot be changed. Section 6, Unit 10 gives a more detailed description of this function.

SETUP PROCEDURES

Reference Points

Reference points are coordinate locations on the machine tool, the workpiece, and in the program.

These points or locations must have a common factor to allow the part to be programmed, setup on the machine tool, and accurately machined. All of the reference points needed in programming the machining center are listed and described in detail in Section 6, Unit 10.

In order to produce accurate parts, it is important that the workpiece, machine spindle, and cutting tools are set up properly. This involves locating the spindle on the edges of the **X and Y** axes, compensating for variations in tool lengths and diameters. For a detailed explanation of these procedures see Unit 7-13.

Locating The Spindle

After a part has been fastened to the table or holding accessory, the center of the spindle should be aligned with the finished **X** and **Y** edges of the part, Figs. 7-14-3 and 7-14-4. See the text and Figs. 7-13-12 and 7-13-13 in Section 7, Unit 13 for a detailed explanation of this procedure.

Tool-Length Compensation

Tool-length compensation is needed because of the wide variations in the lengths of cutting tools used on machining centers. The operator must have some way of informing the machine tool the differences in length of each tool used to machine the part. The most common method of determining the different lengths of tools is by the **Touch Off Method**, see Section 6, Unit 10-9 for more detail.

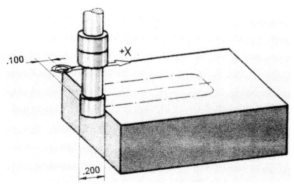

Fig. 7-14-3 Setting the spindle center to zero on the X axis of the part using an edge finder. *(Deckel Maho, Inc.)*

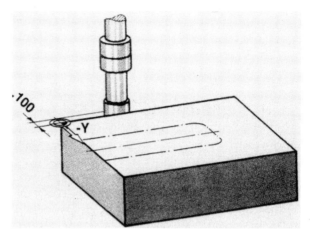

Fig. 7-14-4 Setting the spindle center to zero on the Y axis of the part using an edge finder. *(Deckel Maho, Inc.)*

Cutter-Radius Compensation

Cutter-radius compensation is a feature of the machine control unit that allows the programming of a contour without knowing the exact diameter or radius of the cutter. In programming and machining this feature allows the CNC programmer to write a program without knowing the exact size of the cutter at the time of programming. It also allows the CNC operator to adjust for the size of the cutter in the control system. Adjustment for nominal, undersize, or oversize cutters can be made during actual machining. Cutter-radius compensation can be used for the following conditions:

* unknown cutter size
* adjustment for cutter wear
* adjustment for cutter deflection
* roughing and finishing
* manufacturing variable tolerances

Cutter-Radius Offset

Once the workpiece has been machined, it is checked to see if it is on size, oversize, or undersize. Before inserting a value, be sure the CNC system is designed to accept the compensation on the radius of the cutter and not on the diameter of the cutter, see Unit 7-13. This value is then set as a D-address.

Positive movement of the cutter-radius offset will cause the cutting tool to move away from the machining surface. Negative movement of the cutter-radius offset will cause the cutting tool to move closer to the machining surface.

PROGRAMMING PROCEDURES (G-CODES)

Machining centers are manufactured by many different machine tool builders, and therefore can vary in many ways. The manufacturer of the machine tool may equip the same machine with a different type of machine control unit. This in turn will change the way the machine is to be programmed. Each CNC machine is supplied with a complete operating and programming manual. Before the operator/programmer starts machining, it is his/her responsibility that the correct programming and setup information has been entered into the machine control unit. Due to the number of different machine and control units available it is likely that each machine will vary slightly in the way it is programmed and set up. **Be sure to check the machine manuals before starting to program a job.**

MACHINE PROGRAM

Vertical machining centers are usually the three-axis type or configuration. The **X** axis is the left and right movement of the table, the **Y** axis is the in and out movement of the table, and the **Z** axis is the up or down movement of the spindle or column.

When programming, create the program from the viewpoint of the spindle not of the operator. This means the programmer is looking straight down, perpendicular to the table, when creating the tool path for the program. Most machines have identifications printed on the machine that show the operator the direction of the positive and negative movement of the machine axis. These are operating directions **NOT** programming directions. The programming directions are the exact opposite to these markers on the machine tool.

NOTE: **Program as if the spindle is moving and not the table.**

PREPARATORY FUNCTIONS

The address G identifies a preparatory command. Its one and only function is to preset or to prepare the control system to a certain condition, or a certain mode, or a state of operation. The address G00 presets a rapid motion mode; the address G28 presets the machine zero return command. The term preparatory command indicates its meaning, a G code will prepare the control to accept the programming instructions following the code in a specific way.

Each control system has its own list of available G codes. Many are quite common and can be found on a variety of different controls; several G codes are unique to the particular control system, and may not be considered standard. Because of the nature of machining applications, the list of typical G codes will be different for the milling systems and the turning systems. **Always check the control and machine specifications for the exact meaning of each preparatory G command.**

PREPARATORY FUNCTIONS (G-Codes)

G00	Rapid positioning
G01	Linear interpolation
G02	Circular interpolation clockwise (CW)
G03	Circular interpolation counterclockwise (CCW)
G04	Dwell (as a separate block)
G16	Polar Coordinate Command
G17	**X Y** plane designation
G18	**Z X** plane designation
G19	**Y Z** plane designation
G20	English units of input
G21	Metric units of input
G27	Machine zero position check
G28	Machine zero return (reference point 1)
G29	Return from machine zero
G30	Machine zero return (reference point 2)
G40	Cutter-radius compensation cancel
G41	Cutter-radius compensation - left
G42	Cutter-radius compensation – right
G43	Tool-length compensation - positive
G44	Tool-length compensation - negative
G49	Tool-length compensation cancel

G53	Machine coordinate system
G54	Work coordinate system 1
G80	Fixed cycle cancel
G81	Drilling cycle
G82	Spot-drilling cycle
G84	Right hand threading cycle
G85	Boring cycle
G90	Absolute dimensioning mode
G91	Incremental dimensioning mode
G92	Tool position register
G98	Return to initial level in a fixed cycle
G99	Return to R level in a fixed cycle

Several G codes are special control options—consult the control manual for exact specifications.

MISCELLANEOUS FUNCTIONS (M-CODES)

In addition to the functions relating only to the machine tool, the M functions are used for controlling the execution of the CNC program. Interruption of a program execution requires an M function, during change of a job setup, where one program can call one or more subprograms. Each program then has to have a function, a program call function, the number of repetitions, etc. M functions handle these requirements. The usage of miscellaneous functions can be categorized into two main groups, based on application:

- control of the machine tool functions
- control of the program execution

The following list is the most common miscellaneous functions used for machining centers. Unfortunately, these functions are also the ones that vary the most between machines and control systems. For this reason, **always consult the documentation for the particular machine tool model and its control system.**

TYPICAL APPLICATIONS

Note: the type of activity the M functions perform, regardless of whether that activity relates to the machine tool or the program. Notice the abundance of two-way toggle modes such as ON or OFF, IN or OUT, FORWARD or BACKWARD

APPLICATIONS FOR MILLING

M00	Compulsory program stop
M01	Optional program stop
M02	End of program (reset, no rewind)
M03	Spindle rotation (CW)
M04	Spindle rotation (CCW)
M05	Spindle stop
M06	Automatic tool change
M07	Coolant mist ON
M08	Coolant pump motor ON
M09	Coolant pump motor OFF
M19	Spindle orientation
M30	Program end (reset and rewind)
M48	Feedrate override cancel function OFF (deactivated)
M49	Feedrate override cancel function ON (activated)
M60	Pallet change
M78	B axis clamp
M79	B axis unclamp
M98	Subprogram call
M99	Subprogram end

SAMPLE MACHINING PROJECT

The sample part shown in Fig. 7-14-5 will be used to program some common machining operations performed on a machining center.

```
O1184
        (T01 = 1.000 DIA. END MILL)
N10 G20
N20 G17 G40 G80 T01
N30 M06
N40 G90 G54 G00 X-3.0 Y-3.0 S600 M03
N50 G43 Z0.1 H01 M08
N60 Z-0.6
N70 G41 X-2.0 D01
N80 G01 Y2.0 F15.0
N90 X2.0
N100 Y-2.0
N110 X-3.0
N120 G00 G40 Y-3.0
N130 Z0.1 M09
```

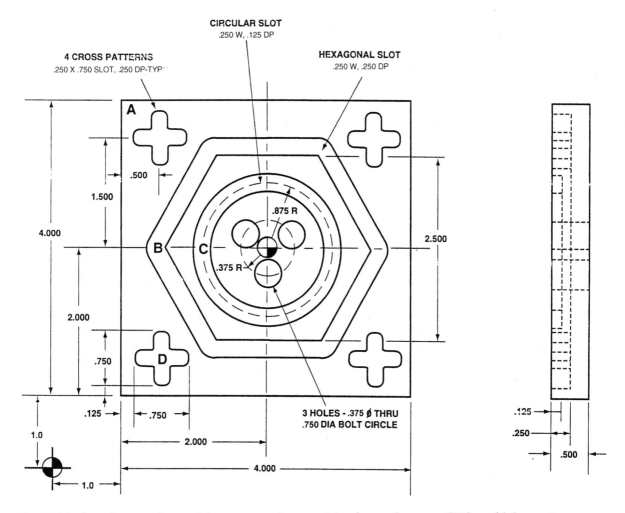

CIRCULAR SLOT
.250 W, .125 DP

4 CROSS PATTERNS
.250 X .750 SLOT, .250 DP-TYP

HEXAGONAL SLOT
.250 W, .250 DP

.500

1.500

.875 R

4.000

2.500

.375 R

2.000

.750

.125   .750

3 HOLES - .375 Ø THRU
.750 DIA BOLT CIRCLE

1.0

2.000

4.000

1.0

.125

.250

.500

Fig. 7-14-5 Sample part print used for programming a variety of operations on a CNC machining center.
(Kelmar Associates)

N140 G91 G28 Z0 M05
N150 M01
     (T02 = .250 DIA. CENTER-CUTTING
          END MILL)
N160 T02
N170 M06
N180 G90 G54 G00 X-1.5877 Y0 S1800 M03
N190 G43 Z0.1 H02 M08
N200 G01 Z-0.25 F5.0
     (CUTTING HEXAGON)
N210 X-0.7939 Y1.375 F8.0
N220 X0.7939
N230 X1.5877 Y0
N240 X0.7939 Y-1.375
N250 X-0.7939

N260 X-1.5877 Y0
N270 G00 Z0.1
     (CUTTING CIRCULAR SLOT)
N280 X-0.875
N290 G01 Z-0.125 F5.0
N300 G02 I0.875 F8.0
N310 G00 Z0.1
     (CUTTING FOUR CROSS PATTERNS)
N320 X-1.5 Y-1.75
N330 G01 Z-0.25 F5.0
N340 Y-1.25
N350 G00 Z0.1
N360 X-1.75 Y-1.5
N370 G01 Z-0.25
N380 X-1.25

N390 G00 Z0.1
N400 Y1.5
N410 G01 Z-0.25
N420 X-1.75
N430 G00 Z0.1
N440 X-1.5 Y1.75
N450 G01 Z-0.25
N460 Y1.25
N470 G00 Z0.1
N480 X1.5
N490 G01 Z-0.25
N500 Y1.75
N510 G00 Z0.1
N520 X1.25 Y1.5
N530 G01 Z-0.25
N540 X1.75
N550 G00 Z0.1
N560 Y-1.5
N570 G01 Z-0.25
N580 X1.25
N590 G00 Z0.1
N600 X1.5 Y-1.25
N610 G01 Z-0.25
N620 Y-1.75
N630 G00 Z0.1 M09
N640 G91 G28 Z0 M05
N650 G90 X-3.0 Y-3.0 (TOOL CHANGE POSI-
       TION)
N660 M01
       (T03 = .500 DIA SPOT DRILL)
N670 T03
N680 M06
N690 G90 G54 G00 X0 Y-0.375 S1400 M03
N700 G43 Z0.1 H03 M08
N710 G99 G82 R0.1 Z-0.2 P250 F6.0
N720 X-0.3248 Y0.1875

N730 X0.3248
N740 G80 Z0.1 M09
N750 G91 G28 Z0 M05
N760 G90 X-3.0 Y-3.0 (TOOL CHANGE POSI-
       TION)
N770 M01
N780 T04 (T04 = .375 DIA DRILL)
N790 M06
N800 G90 G54 G00 X0.3248 Y0.1875 S1250 M03
N810 G43 Z0.1 H04 M08
N820 G99 G81 R0.1 Z-0.663 F8.0
N830 X-0.3248
N840 X0 Y-0.375
N850 G80 Z0.1 M09
N860 G91 G28 Z0 M05
N870 G28 X0 Y0
N880 M30

SUMMARY

- CNC machining centers have become one of the major machine tools used in industry.
- These machines vary in design and application from simple 2 1/2 axis machines to complex 5 axis profiling machines.
- Machining centers can be equipped with automatic tool changers to hold a variety of cutting tools for most machining applications.
- Some machine tools have high-speed machining capabilities, robotic interfaces, and automatic pallet changers to improve the manufacturing process.
- Machines are capable of multi-axis simultaneous cutting tool motion to produce complex contours, angles and surfaces.

KNOWLEDGE REVIEW

Mill Programming

1. What was the first version of a CNC machining center?
2. How many axes were capable of simultaneous cutting tool motion?
3. What would a simple design of machine be similar to?

4. What do all CNC machining centers have in common?

Machine Types

5. Name the three most common machining centers.

6. List the three main categories of machining centers.
7. What are machining centers with the spindle motion designed to move up and down classified as?
8. What are machining centers with the spindle motion designed to move in and out classified as?

Machine Accessories

9. Name five machine operations that can be performed on a machining center?
10. What are five other advanced features that can be added to a machining center?

Machine Axes

11. Name the three main axes of a machining center and state the function of each.
12. What is a fourth axis on a machining center?
13. How is the fourth axis usually designated?
14. When is a machining center considered to be a true four axis machine?

Machine zero point

15. What is the machine zero point?
16. What are three reference points?

Tool length compensation

17. Why is tool length compensation required when setting up the cutting tools?
18. What is the most common method of determining the tool length compensation?

Cutter-radius compensation

19. What is cutter-radius compensation?
20. Name five variables that cutter-radius compensation could be used for.

Cutter-radius offset

21. What effect does positive and negative cutter-radius offset have on the relationship between the cutting tool and the workpiece?
22. Where is the cutter-radius offset value stored in the machine control unit?

Preparatory functions

23. For what purpose are G codes used?
24. Identify the following G codes:
G02, G17, G20, G49, G54.

Miscellaneous functions

25. What functions do M codes control?
26. Identify the following M codes:
M00, M03, M06, M30, M99.

Unit 15
CAD/CAM Mill Programming

CAD/CAM for computer aided design and computer aided manufacturing have revolutionized the way parts are designed and manufactured on CNC machining centers, Fig. 7-15-1. The CAD/CAM software used in a computer system allows the programmer to easily generate complex shapes and surfaces without the use of complicated mathematical formulas. The time needed to design parts, produce mechanical prints, and to machine a workpiece is thus greatly reduced.

A CAD/CAM system is an integrated software program that is capable of designing the workpiece, preparing engineering drawings used to produce CNC programs, and simulating the cutting tool path before machining the part to check for programming errors.

CAD/CAM programs are not only used for milling and lathe applications, but also for a variety of machine tools including wire and ram-type electrical discharge machines, and all types of grinding and drilling machine tools. The list of the machine tools that use CAD/CAM for design and manufacturing would be too numerous to mention. Descriptions of other types of machine tools that use CAD/CAM can be found in trade publications and magazines.

The term CAD/CAM is used in the manufacturing industry and refers to the design of a part (CAD), and the machining of the part (CAM).

CAD software is far more versatile than just designing parts to be machined on CNC machine tools. CAD is used in the automotive and aerospace industries, for landscaping, machine design, and solid modeling for molds and die design and many other applications. CAD is also used by architects to generate working drawings for all the structural and architectural parts of buildings such as walls and roofs. The modules or the entire structure can be displayed in 2D, 2 1/2D and 3D views. CAD images are used to replace the labor-intensive hand perspectives and hand drawings needed when making presentations.

CAM allows the programmer to design the required workpiece in a CAD environment and then use this information to create the cutter path to machine it. The programmer generates the finished CNC program by indicating what cutting tools, machines, and machining operations will be required to produce the part. When this information is put into the software program the finished CNC program is generated. This program can be run to simulate the cutting motion on the computer screen to check for program errors or it can be sent directly to the machine tool to produce the part. It is always recommended that the program be checked for errors before use.

The one-column format used in this CAD/CAM Unit allows the placement of illustrations as close as possible to the corresponding instructions in the text. Students then can immediately and easily see if their computer screen input matches the illustrations in the text, or if a programming error has been made.

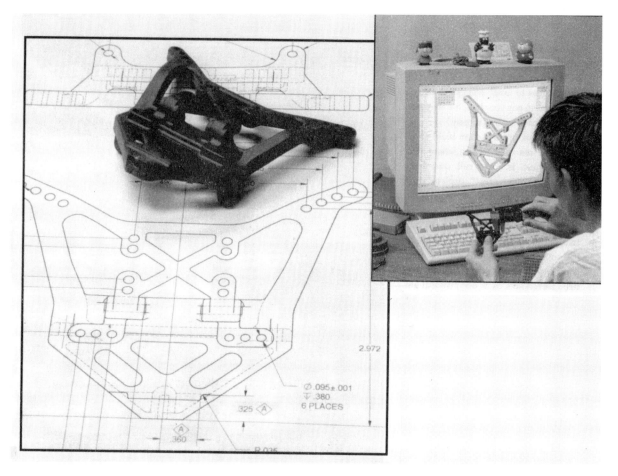

Fig. 7-15-1 CAD/CAM systems allow a part to be designed on a computer screen, to create the CNC program, and to check the tool path. *(Haas Automation, Inc.)*

OBJECTIVES

After completing this unit you should be able to:

1. Design a part using the CAD/CAM software provided.
2. Create a finished program for a CNC machining center.
3. Simulate the cutting tool path.
4. Modify the CNC program to eliminate any errors.

KEY TERMS

CAD/CAM	edit	simulation
CAD images	elements	parameters
drawing screen	mouse	window

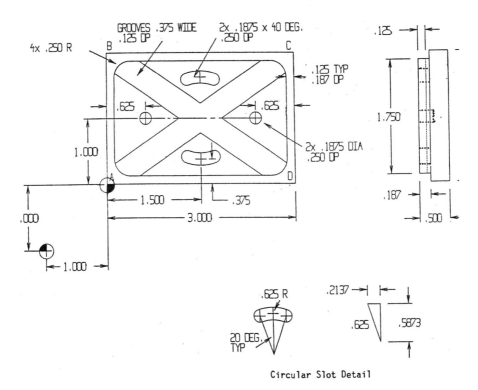

Fig. 7-15-2 **Print of part to be designed.** *(Kelmar Associates)*

DESIGNING A PART

This book includes a CAD/CAM software program that allows a person to design a milling program to machine the part shown in Fig. 7-15-2. Be sure to install the CAD/CAM program on the computer by following the instructions with the software. By carefully following the steps listed it should be easy to program this part.

1) Start the Mill Cam Designer software.
2) The **Material Size** screen will appear. Follow the substeps listed to define the correct settings.
   a) Select **Novamill (inches)** as the Machine Type. To do this, click the down arrow _ shown next to the Machine Type box. A drop-down list of machines will be displayed. Left click on **Novamill (inches)**.
   b) Select **Aluminum** as the Material Type.
   c) Set the part size as follows:
      Height .500
      Width 2.000
      Length 3.000

      To set the part size, highlight the value shown in each window and type the correct size. To highlight, hold down the left mouse button while moving the mouse over the value.

NOTE: To avoid repeating numerous left clicks in an operation, the ➜ sign will be used whenever it is necessary to left click.

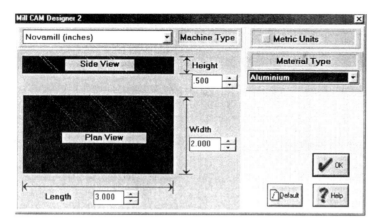

Fig. 7-15-3   Material screen to select machine type, units of measurement, material, and size.

d)   Click (→) the **OK** button when complete.

3) The **Main Drawing screen** will appear. The drawing area equals the stock width and length defined in Step 2.

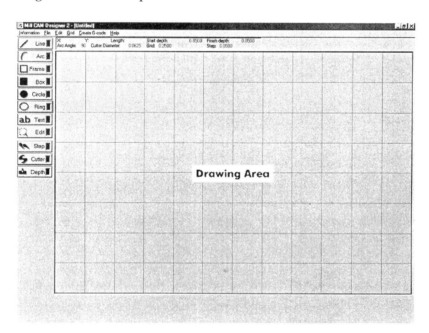

Fig. 7-15-5   Mill Cam Designer drawing screen. *(Denford, Inc.)*

4) To set the **grid size**, from the menu bar, ➔ **Grid** ➔ **0.25 Grid**.

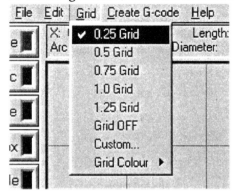

Fig. 7-15-6 Selecting the grid size of the drawing screen. *(Denford, Inc.)*

5) It is important to define the correct cutting parameters before designing the part. From the menu bar, ➔ **Create G-code** ➔ **Set G-code parameters**.

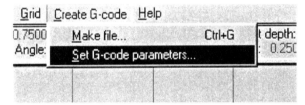

Fig. 7-15-7 Select the correct cutting parameters before designing the part. *(Denford, Inc.)*

6) Two end mills will be used in the manufacture of the part
   • .375 diameter
   • .1875 diameter

   To set the parameters for the .1875 diameter cutter, follow these substeps:
a) ➔ the **0.188″** tick box.

Fig. 7-15-8 Set the depth of cut for the .1875 dia. cutter.

b) Increase the **Cutter Maximum Depth** to 0.300 by highlighting the current value and typing 0.300.

0.300   **Cutter Maximum depth [Inches]**

Fig. 7-15-9 Selecting the .1875 dia. cutter to be used to machine the part. *(Denford, Inc.)*

To set the parameter for the .375 dia. cutter, follow these substeps:
a) ➔ the **0.375"** tick box.

◇ **0.313"**   ◇ 0.5
◆ **0.375"**   ◇ 0.6
◇ **0.438"**   ◇ 0.6
◇ **0.500"**   ◇ 0.7

Fig. 7-15-10   Selecting the .375 dia. cutter to be used. *(Denford, Inc.)*

b) Increase the **Cutter Maximum Depth** to 0.200.

⬚ 0.200   **Cutter Maximum depth [Inches]**

Fig. 7-15-11   Set the depth of cut for the .375 dia. cutter. *(Denford, Inc.)*

c) ➔ the **OK** button when complete.
7) The first element that will be drawn is the outside step. Follow these substeps:
   a) ➔ the **Cutter** button.

   b) ➔ the **0.3750 inches** tick box.

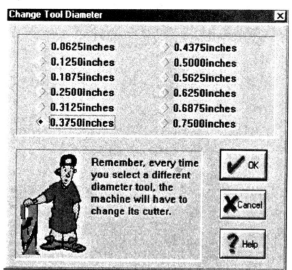

Fig. 7-15-13   Select the cutting tool. *(Denford, Inc.)*

c) ➔ the **OK** button when complete.
d) To set the depth of cut, ➔ the **Depth** button.

➔ the diamond tick box in front of the green square. Change the **Start and End of Line Depths** to read **0.187**, by highlighting the values to the left and right of the green square and typing **0.187**

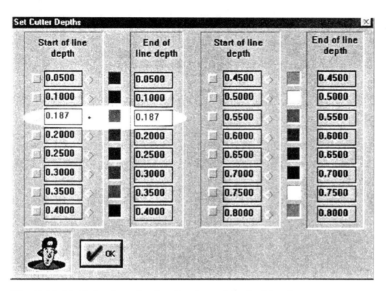

Fig. 7-15-15   Selecting the depth of cut. *(Denford, Inc.)*

e)   → the **OK** button when complete.
f)   → the **Frame** button.

g)   Press the **Enter** key to open the **Manual Data Input** window. The **Frame** will be added to the design by typing the required coordinates.
h)   Change the **X** and **Y** Start Coordinates to **zero**.
Change the End Coordinates to **X3.00, Y2.00**

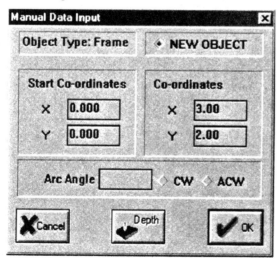

Fig. 7-15-17   **Manual Data Input (MDI) screen to set the XY coordinates of the border.** *(Denford, Inc.)*

→ the **OK** button.

i) The design should now look like Fig. 7-15-18. Notice that the line color (green) matches the depth of cut color selected earlier.

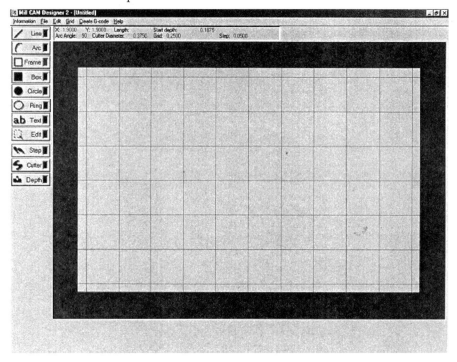

Fig. 7-15-18   Drawing screen with the first element completed. *(Denford, Inc.)*

NOTE: If it is necessary to remove the frame, press the **right mouse button**.

8) Follow these substeps to save the design.
   a)   From the menu bar, ➔ **File** ➔ **Save As**.

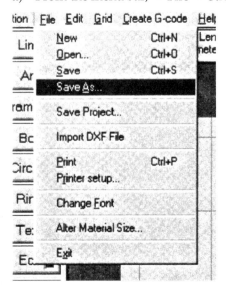

Fig. 7-15-19   File Save screen to save your program. *(Denford, Inc.)*

b) Type a **filename** and ➔ **OK**. A file extension of .mcd will be assigned to the filename.

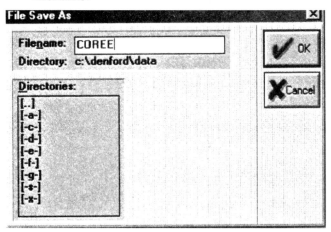

Fig. 7-15-20 **File Save As screen, type in a filename.** *(Denford, Inc.)*

9) The second element to be drawn is a diagonal grove.
   a) The depth of cut must be changed.
      - ➔ the **Depth** button.
      - ➔ the diamond tick in front of the blue square.
      - Change the **Start and End of Line Depth** for the blue selection to **0.125**.

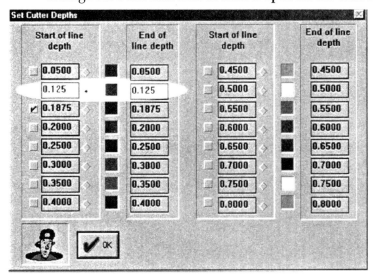

Fig. 7-15-21 **Set cutter depth for second element to be designed.** *(Denford, Inc.)*

NOTE: Check marks show the depths that have been used in the design.

   - ➔ the **OK** button.
   b) ➔ the **Line** button and press **Enter**.

c) Change the **Start Coordinates** to X0 Y0
   Change the **End Coordinates** to X3.000 Y2.000

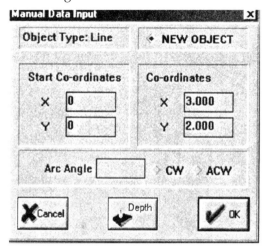

Fig. 7-15-23 MDI screen for the start and end coordinates of the first diagonal line.
(Denford, Inc.)

- → **OK** when complete.

d) The design will now appear as in Fig. 7-15-24. Notice that the line color is blue representing a .1250 in. depth of cut.

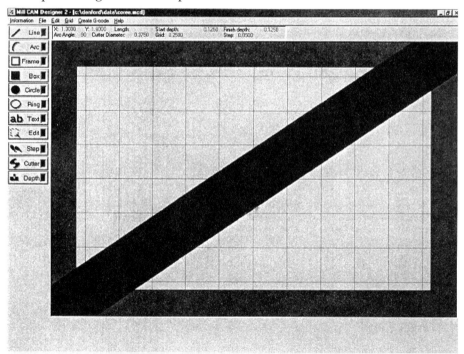

Fig. 7-15-24 Mill Designer screen with the second element completed. *(Denford, Inc.)*

10) The third element is another diagonal groove.
   a) → the **Line** button, Fig. 7-15-22 and press **Enter**.
   b) Change the Start Coordinates to **X0, Y2.000**
      Change the End Coordinates to **X3.000, Y0**

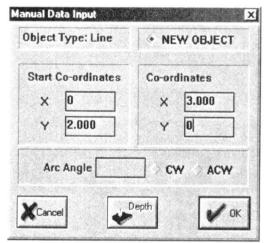

Fig. 7-15-25 MDI screen for the start and end coordinates of the second diagonal line. (*Denford, Inc.*)

- → the **OK** button when complete.

The design should now appear as in Fig. 7-15-26.

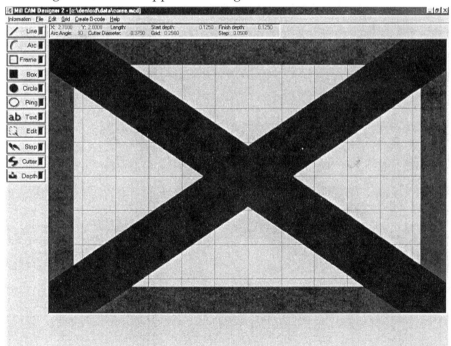

Fig. 7-15-26 Mill Cam Designer screen with the third element completed. (*Denford, Inc.*)

11) The fourth and fifth elements are the drilled holes.
   a) → the **Cutter** button Fig. 7-15-12.
      Select the **0.1875 inches** cutter.

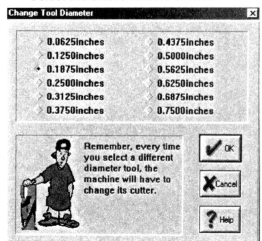

Fig. 7-15-27   Tool-change screen to change the cutting tool for the holes. *(Denford, Inc.)*

   • → OK.
   b) → the **Depth** button.
      Select the **0.2500** depth of cut by clicking the diamond in front of the dark blue square.

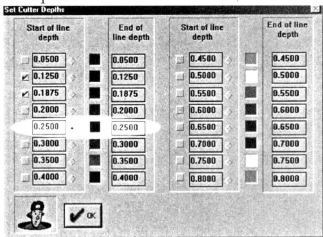

Fig. 7-15-28   Setting cutter depth for the two holes. *(Denford, Inc.)*

c) → the **Line** button and press **Enter**.
    Set the **Start Coordinates** to X.625, Y1.000
    Set the **End Coordinates** to X.625, Y1.000

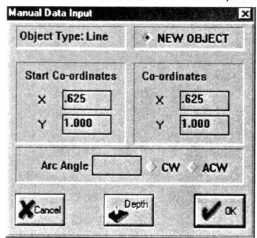

Fig. 7-15-29   MDI screen for the XY coordinates of the first hole location. *(Denford, Inc.)*

- → OK.
d) To create the second hole, → the **Line** button and press **Enter**.
    Set the **Start Coordinates** to X2.375, Y1.000
    Set the **End Coordinates** to X2.375, Y1.000

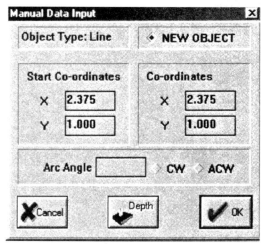

Fig. 7-15-30   MDI screen for the XY coordinates of the second hole location. *(Denford, Inc.)*

- → OK.

The design will now appear as in Fig. 7-15-31

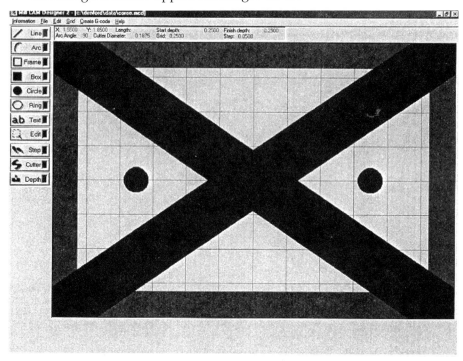

Fig. 7-15-31 Mill Cam Designer screen with the holes completed. *(Denford, Inc.)*

NOTE: If a mistake has been made and it is necessary to remove an element, → the **Edit** button and → the **right mouse button** to cycle through the elements. Once the element to be removed is black in color, → the **left mouse button** once.

Save the design by → **File** → **Save**.

12) The last two elements are the circular slots.
   a) → the **Arc** button and press **Enter**.

b) Type the following coordinates for the upper arc.
   Start Coordinates **X1.2863 Y1.5873**
   End Coordinates **X1.7137 Y1.5873**

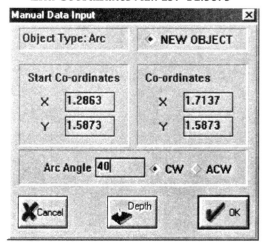

Fig. 7-15-33 MDI screen for start and end coordinates of the first arc. *(Denford, Inc.)*

- → OK.
c) Type the following coordinates to create the lower arc.
   Start Coordinates **X1.2863 Y1.7137**
   End Coordinates **X.4127 Y.4127**

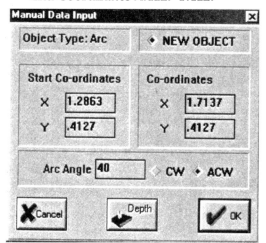

Fig. 7-15-34 MDI screen for the start and end coordinates of the second arc. *(Denford, Inc.)*

- → OK.
  The part is now complete.
  Save the file by → **File** → **Save**.

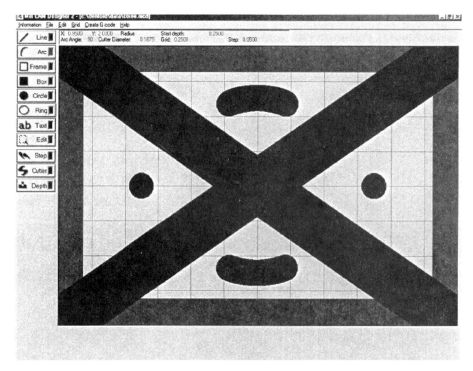

Fig. 7-15-35   Mill Cam Designer screen with the arcs completed.

13) Now that the design is complete, it is time to create the CNC program. CAM software creates the program automatically.
   a) From the menu bar, ➔ **Create G-code** ➔ **Make File**.
   b) Type a unique filename or except the name automatically assigned by the software. AN extension of FNC will be given. ➔ **OK**.

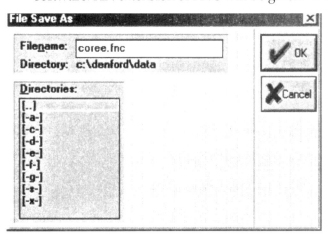

Fig. 7-15-36   File Save screen to save the CNC program. *(Denford, Inc.)*

c) The **G-code Generator** screen will now appear.

Fig. 7-15-37   G-code Generation screen. *(Denford, Inc.)*

* → the **Create G-code** button.

d) The **G-code Timing** window will appear giving an approximate time needed for machining. → the **OK** button.

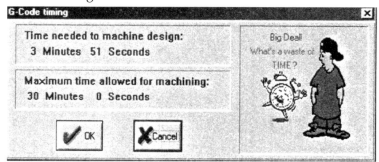

Fig. 7-15-39   G-code Timing window indicating the approximate time required to machine the part. *(Denford, Inc.)*

e) The **Post Processor** will convert the design into CNC code. When the blue line reaches the end of the box, ➔ the **OK** button. The CNC program is now complete.

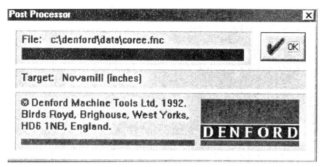

Fig. 7-15-40   Post Processor screen used to convert the design into a CNC program. *(Denford, Inc.)*

14) To exit the software program, ➔ **File** ➔ **Exit**.

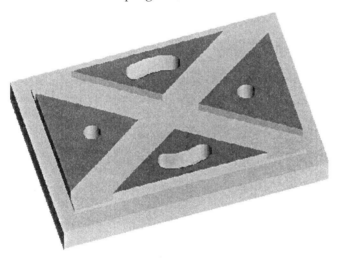

Fig. 7-15-41   Computer simulation of the completed workpiece. *(Denford, Inc.)*

SUMMARY

• CAD/CAM software programs allow the programmer to design the part, select the cutting tools required, and simulate the cutting tool action before starting to machine the part.
• Simulating the cutting action allows the programmer to make corrections to the program, if necessary, before any machining takes place.
• By simulating the cutting action on the computer screen, expensive machine repairs and scrap parts are eliminated.
• CAD/CAM reduces the production time and costs of manufacturing a workpiece.
• Any changes to the design of the part or to the CNC program can be done quickly and accurately.
• CAD/CAM designs and programs can be quickly sent to other manufacturers to produce identical parts.

KNOWLEDGE REVIEW

1. To what do the terms CAD and CAM refer?
2. What are the advantages of CAD/CAM software?
3. List five different machine tools that use CAD/CAM software to program the workpiece.
4. What other industries use CAD software programs?
5. Why is simulation of the cutter path an important feature of the CAM software?

SECTION 8

Simple Programming

CNC turning and machining centers are finding wide acceptance in manufacturing because they increase productivity, reduce manufacturing costs, and provide more flexibility because they can also machine angles and contours, Fig. 8-1. They have often increased productivity from 300 to 500% over conventional machines and reduced operator skill requirements. Operation needs no attention once the CNC program has been checked for accuracy.

Modern CNC turning/chucking centers are equipped with tool turrets that hold numerous different tools for the machining operations required on the part. The machining operations and the tools required for each are included in the CNC program and are changed automatically through the proper commands in the computer program. The time spent on changing tools is thus minimized and preliminary or secondary operations are often reduced or eliminated.

The greatest advantages of CNC turning/chucking centers are:

- Improved part quality due to the accuracy and reliability of the machine tool and the control system.
- Greater productivity because all operations such as tool setups and changes are controlled by the CNC program.
- Shorter lead times result in reduced inventory costs and faster delivery of a product.
- Smaller lot sizes can be machined with cost reductions, especially if complex forms or tapers are required.

Fig. 8-1 A CNC Turning Center. *Courtesy Emco Maier Corp.*

Fig. 8-2 A variety of parts produced on turning and chucking centers.

- Biggest cost savings occur when lot sizes or 10 or more parts are required.
- Cost savings from reduced operator skill requirements, also an operator can look after 2 or more machines at the same time.
- Stored programs can be used at future dates to reproduce a part to the original accuracy.

The continual improvement has resulted in the live tooling accessory that makes it possible to perform operations such as drilling holes at 90° to the centerline, milling slots and keyways, etc., on the part while it is still in the turning center, Fig. 8-2.

Unit 16
Tooling and Workholding Systems

There are a variety of cutting tools that can be used on chucking and turning centers. They can be either conventional outside diameter (OD) or inside diameter (ID) cutting tools. These tools are used for turning, threading, grooving and boring operations, Fig. 8-16-1, and for drilling, center drilling, tapping and reaming. The introduction of motorized tooling on chucking and turning centers made it possible to also perform milling operations on the workpiece while it is still in the machine.

Workholding devices have greatly improved in quality of the chucks available and also in the way they are mounted and removed from the machine tool. Chuck-changing systems are available for mounting and removing chucks and also for changing the jaws in the chuck to accommodate different workpieces.

OBJECTIVES

After completing this unit, you should be able to:

1. Identify the different cutting tools used on turning and chucking centers.
2. Know and understand the importance of workholding devices used for turning.
3. Know and understand the purpose of tool-monitoring systems and in-process gaging.

KEY TERMS

centrifugal force	tool coatings
clamping pressure	tool monitoring
high-velocity tooling	tooling systems
in-process gaging	workholding devices

CUTTING TOOLS

A variety of types and styles of turning tools and indexable-insert tooling is available for chucking and turning centers. Indexable-type tooling allows the operator to change or index inserts at the machine tool instead of removing the cutting tool for resharpening and replacing it with another tool.

Fig. 8-16-1 A variety of cutting tools used on chucking and turning centers. *(Kennametal Inc.)*

Coatings

Thin wear-resistant coatings, applied to cutting tools, have resulted in increased tool life, increased productivity, and freer chip flow. The coating acts as a lubricant reducing cutting forces, heat, and wear on the cutting tool, permitting higher spindle speeds and improved surface finishes.

Types of coatings

The use of hard, wear-resistant coatings of carbides, nitrides, and oxides to cutting tools has improved their performance. The main benefits they offer are increased productivity, longer tool life, and reduced machining costs. The most common tool coatings are as follows:

- Titanium Nitride (TiN) a gold-colored coating
- Titanium Carbonitride (TiCN) a blue-gray colored coating
- Chromium Nitride (CrN) a silver-gray colored coating

- Chromium Carbide (CrC) a silver-gray colored coating
- Titanium Aluminum Nitride (TiAIN) a violet-gray colored coating
- Tungsten Carbide/Carbon (WC/C) a black-gray colored coating

For a list of properties and applications for thin wear-resistant coatings refer to Table 8-16-1.

Carbide Cutting Tools

Carbide turning tools are available in basically two types, brazed-tip (cemented) tools, Fig. 8-16-2, and indexable insert tools, Fig. 8-16-3. Brazed-tip carbide tools have a carbide insert that has been brazed onto a steel shank. They are available in a wide variety of sizes, shapes, and styles as well as grades of carbide for specific applications. The two main groups in which carbide is classified are tungsten carbide and titanium and/or tantalum carbide (crater-resistant grades).

Carbide Insert Identification

The American Standards Association has developed a system to quickly and accurately identify indexable inserts that most carbide insert manufacturers use. For a list of insert types and applications refer to Table 8-16-2.

Carbide Inserts

Because of the tremendous variety of carbide

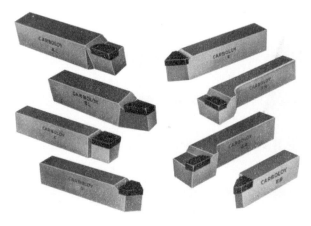

Fig. 8-16-2 A variety of brazed carbide cutting tools. *(Carboloy Inc.)*

	Titanium Nitride	Titanium Carbonitride	Chromium Nitride	Chromium Carbide	Titanium Aluminum Nitride		Tungsten Carbide/ Carbon
	Balinit® A	Balinit® B	Balinit® D	Balinit® CAST	Balinit® FUTURA	Balinit® X.TREME	Balinit® C
Coating Material	TiN	TiCN	CrN	CrC	TiAlN	TiAlN	WC/C
Microhardness (HV 0.05)	2300	3000	1750	1850	3000	3500	1000
Coefficient of Friction Against Steel (Dry)	0.4	0.4	0.5	0.4	0.4	0.4	0.1 - 0.2
Coating Thickness [μm]	1-4	1-4	1-6+	1-6+	1-5	1-3	1-4
Max. Working Temperature	600°C 1100°F	400°C 750°F	700°C 1300°F	700°C 1300°F	800°C 1470°F	800°C 1470°F	300°C 570°F
Coating Color	Gold	Blue-Gray	Silver-Gray	Silver-Gray	Violet-Gray	Violet-Gray	Black-Gray
Key Characteristics	Good general purpose	High hardness, good wear resistance, enhanced toughness	Good adhesion, good corrosion and oxidation resistance	Good adhesion, good corrosion and oxidation resistance	Excellent oxidation resistance	Excellent oxidation resistance	Lubricity Low tendency for adhesive wear
Primary Applications	▪ Machining of iron-based materials ▪ Metal forming ▪ Plastic molding	▪ For mechanically stressed cutting edges - machining difficult-to-machine steel alloys and high speed cutting where moderate temperatures are generated at the cutting edge. ▪ Metal forming	▪ Machining copper ▪ Metal forming ▪ Plastic molding	▪ Aluminum and Magnesium die casting	▪ A multi-layer coating designed for a wide range of carbide, cermet and high speed steel tooling. Excellent for machining cast iron, stainless steel, nickel-based high temperature alloys and titanium alloys. Designed for high speed and semi-dry or dry machining operations.	▪ A specialized coating designed specifically for use on carbide end mills for the machining of hardened steel workpieces. Designed for high speed and semi-dry or dry machining operations.	▪ Precision components ▪ Dry machining ▪ Plastic molding

Table 8-16-1 **Applications and properties of various thin film wear resistant coatings.** *(Balzers Tool Coating, Inc.)*

inserts available, Fig. 8-16-4, it is important that the correct grade and shape be selected for the type of material and machining application required, Table 8-16-3.

Two main considerations when selecting the proper cutting tool insert are:

- Is the insert capable of cutting the required contours?
- Does the insert have sufficient strength to complete the cut?

There is a complex relationship between cutting condition, speed, feed, and tool life. Some of the factors that will affect tool life are:

coolant	insert grade
depth of cut	tool-nose radius
insert geometry	workpiece material

When programming and setting up the machine tool the programmer/operator should look at all the variables relating to the machining operation as if it were a balance scale. If one of the conditions is removed, added, or changed, then another must be changed to keep all conditions in balance. If this balancing is done, a successful machining operation will result. For example, increase the spindle speed and decrease the feed rate to maintain balance.

ANSI Insert Nomenclature

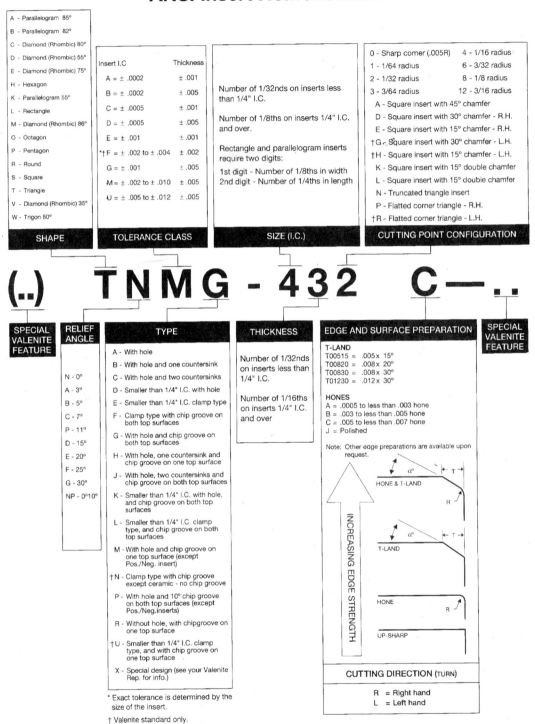

SHAPE

A - Parallelogram 85°
B - Parallelogram 82°
C - Diamond (Rhombic) 80°
D - Diamond (Rhombic) 55°
E - Diamond (Rhombic) 75°
H - Hexagon
K - Parallelogram 55°
L - Rectangle
M - Diamond (Rhombic) 86°
O - Octagon
P - Pentagon
R - Round
S - Square
T - Triangle
V - Diamond (Rhombic) 35°
W - Trigon 80°

TOLERANCE CLASS

Insert I.C	Thickness
A = ± .0002	± .001
B = ± .0002	± .005
C = ± .0005	± .001
D = ± .0005	± .005
E = ± .001	± .001
*†F = ± .002 to ± .004	± .002
G = ± .001	± .005
M = ± .002 to ± .010	± .005
U = ± .005 to ± .012	± .005

SIZE (I.C.)

Number of 1/32nds on inserts less than 1/4" I.C.

Number of 1/8ths on inserts 1/4" I.C. and over.

Rectangle and parallelogram inserts require two digits:
1st digit – Number of 1/8ths in width
2nd digit – Number of 1/4ths in length

CUTTING POINT CONFIGURATION

0 - Sharp corner (.005R)   4 - 1/16 radius
1 - 1/64 radius   6 - 3/32 radius
2 - 1/32 radius   8 - 1/8 radius
3 - 3/64 radius   12 - 3/16 radius
A - Square insert with 45° chamfer
D - Square insert with 30° chamfer - R.H.
E - Square insert with 15° chamfer - R.H.
†G - Square insert with 30° chamfer - L.H.
†H - Square insert with 15° chamfer - L.H.
K - Square insert with 15° double chamfer
L - Square insert with 15° double chamfer
N - Truncated triangle insert
P - Flatted corner triangle - R.H.
†R - Flatted corner triangle - L.H.

(..) T N M G - 4 3 2   C—..

SPECIAL VALENITE FEATURE

RELIEF ANGLE

N - 0°
A - 3°
B - 5°
C - 7°
P - 11°
D - 15°
E - 20°
F - 25°
G - 30°
NP - 0°10°

TYPE

A - With hole
B - With hole and one countersink
C - With hole and two countersinks
D - Smaller than 1/4" I.C. with hole
E - Smaller than 1/4" I.C. clamp type
F - Clamp type with chip groove on both top surfaces
G - With hole and chip groove on both top surfaces
H - With hole, one countersink and chip groove on one top surface
J - With hole, two countersinks and chip groove on both top surfaces
K - Smaller than 1/4" I.C. with hole, and chip groove on both top surfaces
L - Smaller than 1/4" I.C. clamp type, and chip groove on both top surfaces
M - With hole and chip groove on one top surface (except Pos./Neg. insert)
†N - Clamp type with chip groove except ceramic - no chip groove
P - With hole and 10° chip groove on both top surfaces (except Pos./Neg.inserts)
R - Without hole, with chipgroove on one top surface
†U - Smaller than 1/4" I.C. clamp type, and with chip groove on one top surface
X - Special design (see your Valenite Rep. for info.)

* Exact tolerance is determined by the size of the insert.

† Valenite standard only.

THICKNESS

Number of 1/32nds on inserts less than 1/4" I.C.

Number of 1/16ths on inserts 1/4" I.C. and over

EDGE AND SURFACE PREPARATION

T-LAND
T00515 = .005 x 15°
T00820 = .008 x 20°
T00830 = .008 x 30°
T01230 = .012 x 30°

HONES
A = .0005 to less than .003 hone
B = .003 to less than .005 hone
C = .005 to less than .007 hone
J = Polished

Note: Other edge preparations are available upon request.

INCREASING EDGE STRENGTH

HONE & T-LAND
T-LAND
HONE
UP-SHARP

CUTTING DIRECTION (TURN)

R = Right hand
L = Left hand

SPECIAL VALENITE FEATURE

Table 8-16-2   Cemented-carbide insert identification system. *(Carboloy Inc.)*

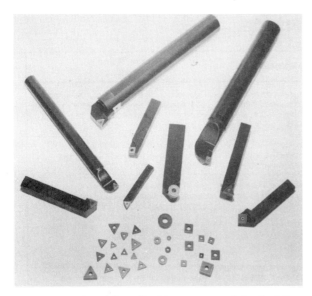

Fig. 8-16-3 Special toolholders are used for holding indexable insert tools. *(Kennametal Inc.)*

Fig. 8-16-4 Carbide-insert cutting tools increase productivity and reduce machining time. *(Carboloy Inc)*

Table 8-16-4 lists some of the many conditions that affect the turning operation, and provides some helpful hints to maintain a balance with regard to the cutting tool.

Coated-Carbide Inserts

Coated-carbide inserts have a thin layer of wear resistant titanium, titanium carbide or aluminum oxide. This coating provides improved cutting-edge wear resistance, increasing the life of the insert and allowing higher cutting speeds. Titanium-coated inserts give the greatest wear resistance at speeds below 500 sf/min., Table 8-16-3.

Ceramic Inserts

Ceramic inserts, Fig. 8-16-5, are manufactured from a heat-resistant material. The most popular material used for ceramic inserts is aluminum oxide. Other materials such as titanium oxide or titanium carbide may be added with the ceramic to produce cutting tools for specific applications. Ceramic tools permit increased tool life, improved surface finish, and higher cutting speeds than carbide cutting tools. These inserts are not as strong as carbide and should be used for applications with uninterrupted cuts and low shock, Table 8-16-5.

Cermet Inserts

Cermet inserts are cutting tool inserts made from ceramic and metal, usually aluminum oxide, titanium carbide, and zirconium oxide. Cermet inserts are used for machining hardened steels at high temperatures, Table, 8-16-6. These inserts provide increased tool life, and improved surface finish that could eliminate some cylindrical grinding operations.

Fig. 8-16-5 Ceramic inserts used for improved surface finish to close tolerances. *(Kennametal Inc.)*

Uncoated Carbide

Valenite Grade	ISO Class	Industry Class	Application	Materials	Working Methods & Conditions
VC2	M10-20 K10-20	C2	Turning, Boring & Milling	Cast iron, copper, brass, non-ferrous alloys, high temperature exotics, stone and plastics	General purpose grade of high toughness and resistance against flank wear at low to medium cutting speeds.
VC3	K01-05	C3,C4	Precision Turning, Boring & Milling	Cast iron, aluminum, high temperature exotics and non-ferrous materials	Wear-resistant grade for finishing cuts, low to medium feed rates under rigid conditions.
VC5	P20-30 M20-40	C5	Turning, Boring & Milling	Steel, cast steel, malleable cast iron, 400/500 series stainless steels	General purpose grade covering a wide range of applications, low to medium cutting speeds, high feeds and depths of cut. Has good deformation resistance.
VC7	P05-15	C7	Turning, Boring, Grooving & Threading	Steel, cast steel, malleable cast iron, 400/500 series stainless steels	Light roughing to finishing at low to moderate feeds. Good crater and deformation resistance.
VC8	P01-10	C8	Precision Turning & Boring	Steel, cast steel, malleable cast iron, 400/500 series stainless steels	High-speed finishing grade with best thermal and deformation resistance.
VC27	P15-30 M15-30 K20-30	C2	Turning & Milling	Steel, cast steel, alloyed cast irons, cast alloys, exotics	General purpose fine grain grade with improved toughness and wear resistance for turning and milling.
VC28	M20-30 K15-30	C2	Milling	Cast and alloy irons	General purpose grade for roughing to finishing in cast irons.
VC29	M10-20 K10-20	C2,C3	Turning, Boring & Milling	Stainless steels, irons, exotics, and non-ferrous metals	Fine grain grade for finishing of exotic irons and non-ferrous metals.
VC35M*	P20-35	C5	Milling	Carbon, alloy steel and stainless steel	General purpose steel milling grade for moderate roughing to finishing.
VC101	M30-40 K30-40	C1	Turning, Boring & Milling	Iron, 200/300 stainless steel and exotics	Fine grain heavy duty grade for roughing at low to moderate speeds.
VC111	M30-45 K30-45	C1	Turning, Boring, Milling & Drilling	Iron, stainless steel and exotics	Heavy duty grade for roughing where low speeds and heavy impact are present.
VC121*	K05-30	C2	V-Cut Plus Cut off & grooving	Cast iron, aluminum, non-ferrous alloys, exotics	Medium cutting speeds and feeds.
VC135*	P25-50	C5	V-Cut Plus Cut off & grooving	Steel, stainless steel	Medium speeds, heavy feeds in unfavorable conditions.

Table 8-16-3 Coated and uncoated carbide grades for metal cutting applications. *(Valenite Inc.)*

Cubic Boron Nitride

Cubic boron nitride (CBN) is second only to diamond in hardness. These inserts are manufactured by bonding a layer of polycrystalline cubic boron nitride (PCBN) to a cemented-carbide substrate (base), Fig. 8-16-6. These inserts have good shock resistance, high-wear resistance, and good cutting-edge life. PCBN inserts are used to machine hardened ferrous metals and alloys.

Diamond Inserts

Diamond inserts are made from tiny manufactured diamonds bonded to a carbide substrate, Fig.8-16-6. They are generally manufactured into polycrystalline diamond (PCD) inserts that are used to machine nonferrous metals, and abrasive non-metallic materials. PCD inserts offer increased wear and shock resistance as well as greatly increased cutting speeds and improved

Coated Carbides

Valenite Grade	ISO Class	Industry Class	Application	Materials	Working Methods & Conditions
VN5	P10-25 M15-20	C5-C7	Turning, & Boring	Steel, cast steel, malleable cast iron, stainless steel	A TiN coated grade for roughing and finishing. Has excellent crater and deformation resistance.
VN8	P10-30 M20-30 K10-30	C2 C7	Turning, Boring, Milling , Threading & Grooving	Cast iron, steels, 300 and 400 series stainless steels, PH stainless steels	A very well balanced TiN coated grade suitable for a broad range of applications. Has outstanding crater and impact resistance at low to high speeds.
V01	P01-30 M10-30 K01-20	C2 C8	Turning, Boring & Milling	Cast iron, stainless steels, alloyed steels, carbon steels	A composite ceramic coated grade providing maximum resistance to built up edge. Suitable for operations ranging from roughing to finishing at medium to high speeds.
V05	P01-30 M10-30 K01-20	C2, C8	Turning, Boring & Milling	Cast iron, alloy steel, stainless steel, & carbon steel	A composite ceramic coated grade optimized for wear resistance with good impact and built up edge resistance. A good choice for machining difficult materials.
V1N	P30-45 M30-40 K25-45	C1,C2, C5	Milling, Turning, Grooving & Threading	Cast iron, steels, high temperature exotics, 300 and 400 series stainless steels, and PH stainless steels	A TiN coated heavy duty grade used in severe roughing and interrupted cuts at slow speeds.
V88	P05-30 M10-30 K05-30	C2 C7	Turning, Boring, & Milling	Cast iron, steel, and alloy steel	A TiC coated grade with excellent flank wear resistance for use in applications where abrasive wear is the primary failure mode.
VX8	P15-30 M15-30 K15-30	C2, C5	Turning, Boring, Milling & Threading	Cast iron, steels, high temperature exotics, and stainless steels	A TiC and TiN coated grade for moderate to heavy cuts with medium to heavy feeds. Optimized for flank wear resistance.

Table 8-16-3   Continued

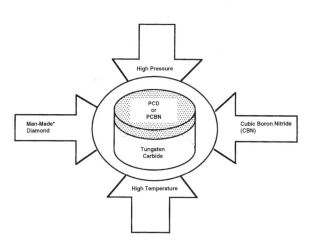

Fig. 8-16-6 CBN or Diamond layer bonded to a tungsten carbide substrate. *(GE Superabrasives)*

surface finishes. Diamond tool life can be 100 times greater than the life of a carbide insert.

TOOLING SYSTEMS

Tooling systems for turning and chucking centers may vary with individual manufacturer's specifications. It is important to remember that the success of any turning operation will depend on the accuracy of the tooling system and the cutting tools being used. A typical tooling system, Fig. 8-16-7, consists of toolholders, boring bar holders, facing and turning holders, and drill sockets.

Preset Tools

Preset tooling is cutting tools that are preset to exact dimensions for a specific machining opera-

problem/concern	possible causes and areas of investigation	speeds (sfm)	feed	depth of cut	grade	coolant	rake angle	edge preparation	material (type/condition)	center height	geometry (insert)	insert finish	insert thickness	nose radius	lead angle	holder (type/condition)	machine condition	chip flow direction	horsepower	excessive overhang	gibs (worn/out of adjust.)	spindle bearings	turret	head stock	machine level	machine anchored	workholding	leadscrew	rigidity	chatter
unacceptable chips	stringer/ribbons (light silver color)	P↑	P↑	•		•	•	•	•		P			•	•			•													
unacceptable chips	corrugated/tight (dark blue or black color)	•	P↓			•	•	•	•		P			•	•																
workpiece concerns	finish/rms tolerance	P	P		•	•	•	•	•	•		P			•	•															
workpiece concerns	interrupted cuts	P↑	P↓	P↓		•	•	•	•		•		•	•	↑				•	•	•	•				•	•			•	•
machine concerns	areas of investigation	•	•			•			•							•			•	•	•	•				•		•	P		
insert failure modes	edge wear	P	P	•	P	•		•																							
insert failure modes	heat deformation (up-set)	P↓	P↓	P↓	•	•								•	•																
insert failure modes	thermal cracking	•	•	•	P	P					•			•																	
insert failure modes	crater	P↓	P↓		•	•	•				•																				
insert failure modes	chipping	•	•		P	•	•	P	•	•		•	•	•															•	•	•
insert failure modes	depth-of-cut notching	•	•		•	•	•	P		•			P																		
insert failure modes	built-up edge	P↑	P↑		•	•	•		P	•	•																				
insert failure modes	catastrophic breakage	•	•	•	•	•	•	•	•	•	•	•		•			•										•			P	P

↑↓ Arrows indicate direction of adjustment
"P" indicates areas of primary investigation

A94-103(10)E4

Table 8-16-4 Possible causes of cutting tool faults during machining. (*Kennametal Inc.*)

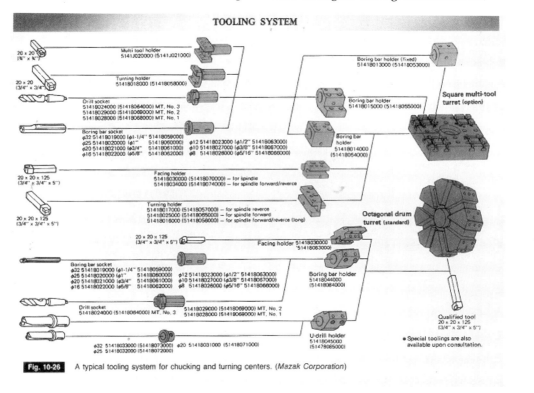

Fig. 10-26 A typical tooling system for chucking and turning centers. (*Mazak Corporation*)

Fig. 8-16-7 A typical tooling system for chucking and turning centers. (*Mazak Corporation*)

TABLE 8-16-5 Recommended Cutting Speeds for Ceramic Cutting Tools

Workpiece Material	Material Condition or Type	Roughing Cut		Finishing Cut		Recommended Tool Geometry (Type of Rake Angle)	Recommended Coolant
		Depth >.062 in. Feed .015–.030 in.	Depth >1.6 mm Feed .4–0.75 mm	Depth <.062 in. Feed .010 in.	Depth <1.6 mm Feed 0.25 mm		
Carbon and tool steels	Annealed	300–1500	90–455	600–2000	185–610	Neg.	None
	Heat-treated	300–1000	90–305	500–1200	150–365	Neg.	
	Scale	300–800	90–245			Neg. honed edge	
Alloy steels	Annealed	300–800	90–245	400–1400	120–425	Neg.	None
	Heat-treated	300–800	90–245	300–1000	90–305	Neg. honed edge	
	Scale	300–600	90–185			Neg. honed edge	
High-speed steel	Annealed	100–800	30–245	100–1000	30–305	Neg.	None
	Heat-treated	100–600	30–185	100–600	30–385	Neg. honed edge	
	Scale	100–600	30–185			Neg. honed edge	
Stainless steel	300 series	300–1000	90–305	400–1200	120–365	Pos. and neg.	Sulfur base oil
	400 series	300–1000	90–305	400–1200	120–365	Neg.	
Cast iron	Gray iron	200–800	60–245	200–2000	60–610	Pos. and neg.	None
	Pearlitic	200–800	60–245	200–2000	60–610	Neg.	
	Ductile	200–600	60–185	200–1400	60–427	Neg.	
	Chilled	100–600	30–185	200–1400	60–427	Neg. honed edge	
Copper and alloys	Pure	400–800	120–245	600–1400	185–425	Pos. and neg.	Mist coolant
	Brass	400–800	120–245	600–1200	185–365	Pos. and neg.	Mist coolant
	Bronze	150–800	45–245	150–1000	45–305	Pos. and neg.	Mist coolant
Aluminum alloys*		400–2000	120–610	600–3000	185–915	Pos.	None
Magnesium alloys		800–10,000	245–3050	800–10,000	245–3050	Pos.	None
Non-metallics	Green ceramics	300–600	90–185	500–1000	150–305	Pos.	None
	Rubber	300–1000	90–305	400–1200	120–365	Pos.	None
	Carbon	400–1000	120–305	600–2000	185–610	Pos.	None
Plastics		300–1000	90–305	400–1200	120–365	Pos.	None

*Alumina-based cutting rods have a tendency to develop a built-up cutting edge on certain aluminum alloys.

tion as specified by the programmer. Presetting of tools is done in a presetting device, Fig. 8-16-8, or in the machine tool, Fig. 8-16-9. With tool compensation capabilities and qualified tooling the need for preset tooling has been greatly reduced.

Qualified Tooling

Qualified tooling guarantees that the position of the cutting edges will be within specified tolerances. Both the tool inserts and the toolholder are qualified to manufacturers' tight tolerances from specific datums: the end, front and back surfaces of the toolholder. Qualified tooling eliminates the need for presetting and measuring cutting tools.

TOOL MONITORING

When machining is done on a turning center, tool wear and breakage require the continuous attention of the operator. A tool-monitoring system, Fig. 8-16-10, can be substituted for the skilled operator's eyes and ears, signaling in a variety of ways the need to replace tools that are worn and broken. There are many types of tool-monitoring

Fig. 8-16-8 A gage for presetting cutting tools for a chucking or turning center.
(KPT Kaiser Precision Tooling, Inc.)

Fig. 8-16-9 A tool setter calculates the tool offset value.
(Cincinnati Machine, A UNOVA Co.)

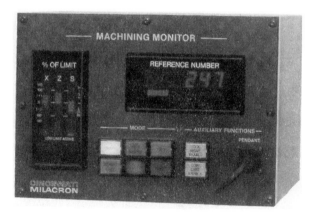

Fig. 8-16-10 A tool monitoring system, informs the operator when a cutting tool is worn and needs replacing. *(Cincinnati Machine, A UNOVA Co.)*

systems available, and the way they detect tool wear varies with the manufacturer. Given that a worn or dull tool requires more power to machine a workpiece than a sharp tool, the most common method of determining wear is from the power or force it takes to drive the cutting process. The tool monitoring system measures the load on the main spindle drive motor in two stages:

1. When the machine is set up and ready, the normal machining cycle is run with new tools and feeds at 100% percent of programmed rate.
2. When the workpiece is completed, a second machining cycle is run, this time without contacting the workpiece.

From these two cycles, the monitoring system can calculate the net machining forces and torque

Material	Hardness (Brinell)	Cutting Speed	Feed	Depth of Cut
Cast irons	100–250	200–1200 sf/min (60–366 m/min)	.002–.016 in. (0.05–0.4 mm)	.187–.250 in. (4.74–6.35 mm)
Steel, carbon	100–250	160–1200 sf/min (48–366 m/min)	.002–.016 in. (0.05–0.4 mm)	.200–.300 in. (5.08–7.62 mm)
Steels, alloys and stainless	250–400	150–1000 sf/min (46–305 m/min)	.002–.016 in. (0.05–0.4 mm)	.187–.300 in. (4.74–7.62 mm)

Table 8-16-5 Recommended cutting speeds for ceramic cutting tools.

Fig. 8-16-11 In-process gaging measures the workpiece and can compensate for tool wear during a machining operation. *(Cincinnati Machine, A UNOVA Co.)*

Fig. 8-16-12 Self-centering chucks are recommended for bar stock, forgings and castings. *(Cincinnati Machine, A UNOVA Co.)*

for every portion of the part program where monitoring is desired. Once the limits have been set, the monitoring system will signal the operator when machining forces and torque exceed acceptable limits and, in some cases, automatically reduce the speed or feed to compensate for the dullness of the cutting tool.

Some methods of detecting tool wear and breakage are through electrical resistance, measuring heat, sound, optical magnification, and vibration. Regardless of the system of detection used, the tool-wear monitoring system provides benefits such as:

- broken tool detection
- machine protection
- improved productivity
- reduced operator attention
- worn tool detection
- In-Process Gaging

An in-process gaging system, Fig. 8-16-11, is a way of monitoring what is happening to the workpiece and tools during the machining oper-

ations. It can also be used to compensate for tool wear and thermal growth, determining tool offsets, locating workpieces, and in datum location and inspection.

The **probe** is a precision electronic surface-sensing device that sends a signal to the machine control system when it is deflected in any direction upon contact with the workpiece or the tool, Fig. 8-16-12. In-cycle gages are omnidirectional, meaning that the sensing probes will detect any movement in the **X or Z** direction. Once the probe stylus is deflected, a signal is sent to the control where the data can be acted on.

On chucking and turning centers, the probes can be mounted either in a toolholder or in a turret. The probes can be selected the same way a cutting tool is selected, by calling it up in the machining program. Probes are used to check a tool for wear and make the appropriate compensation, or to check a part for size between machining operations. The use of in-process gaging helps to reduce operator errors in setup and allows for inspection of fully machined parts in the machine.

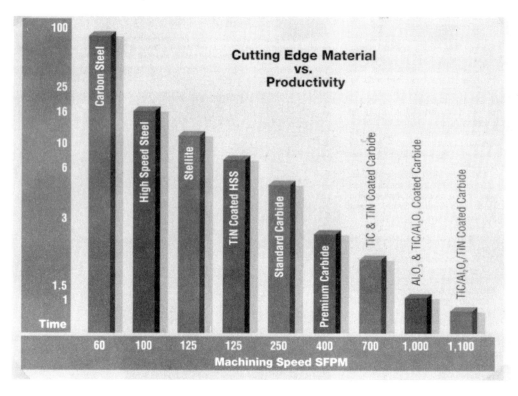

Table 8-16-6 Recommended cutting speeds and feeds for cermet cutting tools

SPEEDS AND FEEDS

There are charts and tables available as guidelines for setting surface speed, feed, and depth of cut, Table 8-16-5. When choosing a surface speed, look at the specific conditions that relate to the workpiece being machined. Interrupted cuts, large depth of cuts, long continuous cuts, surface scale, high feed rates, no coolant, rigidity of setup or workpiece can call for a reduction in the recommended surface speed. Whereas uninterrupted cuts, light feed and depth of cuts, short length of cuts, smooth prefinished materials, flood coolant, and rigid setup allow for recommended surface speeds while maintaining an acceptable tool life.

The correct feed should be used for all types of cutting tools and workpiece materials. Too slow a feed rate results in chip control problems, reduced cutter life, and reduced productivity. Too fast a feed rate can result in insert chipping, breakage and reduced cutter life.

Table 8-16-7 Recommendations for Setting Feeds and Speeds

Use More Feed For:	Use Less Feed For:
eaasy-to-machine materials	deep cutting slots
controlling continuous chips	better finishes
light cuts	reducing edge chipping
reducing chatter	milling thin wall parts
Use Higher Speeds For:	**Use Lower Speeds For:**
soft materials	hard, abrasive materials
better finishes	heavy cuts/rigid set-ups
light cuts	high nickel or manganese
excessive edge chipping	sandy castings
small diameter tools	reducing tool wear

Note: The cutting speed tables listed can be used to calculate spindle speeds; however, such speeds are approximate and should be adjusted for the type and condition of the machine, dimensions of the workpiece, type of machining operation, type of cutting tool and material being machined.

HIGH-VELOCITY TOOLING

High-velocity machining requires new cutting edge materials that are capable of removing material at higher cutting speeds and maintaining their cutting edges longer. Greater amounts of metal are thus removed per unit of time (called Metal Removal Rate (MRR)).

Table 8-16-7 shows how cutting tool materials and machining speeds (sf/min) increased as cutting tool materials improved from carbon steel to the triple coated (TiC/Al$_2$O$_3$) carbides. The table compares various cutting tool-edge materials with the relative time required to complete the same job. As the time to complete a job decreases, a corresponding increase in productivity occurs.

Key factors in high-speed machining are the characteristics of the cutting-tool materials:

- They must be strong enough to withstand the force (pressure) created during the machining operation.
- They must be chemically inert or non-reactive with the workpiece material. Machining steel with a diamond-cutting tool will cause a chemical reaction that can quickly destroy an expensive cutting tool.
- They must maintain their hot hardness at the high temperatures created during the metal-removal process.

WORKHOLDING DEVICES

The most common workholding device on turning and chucking centers is a chuck. There are a wide variety of chucks available, such as self-centering, counter-centrifugal, and collets to suit various workpieces and machining conditions. The demand for heavier metal removal rates and higher spindle speeds has created a need for high-performance chucks. Along with these performance requirements is the need for better and more secure gripping of the workpiece, quick-change jaws, and chucks that can handle different size workpieces, etc.

Self-centering chucks

Self-centering chucks are designed to move all jaws equally and simultaneously to center the part in the chuck. Self-centering chucks, Fig. 8-16-12, normally have higher gripping forces and are more accurate than other chuck types. These chucks are recommended for bar stock, forgings, castings, or turned parts that are located from the gripping diameter. The self-centering chuck is front-actuated and is arranged to hold collet pads in the master jaws for bar stock operations.

Counter-centrifugal chucks

The counter-centrifugal chuck is one way manufacturers have met the need to better grip the workpiece at high speeds. The counter-centrifugal chuck reduces the centrifugal force developed by the high r/min, using counterweights that pivot so that the centrifugal force tends to increase the gripping pressure, thus offsetting the outward forces developed by centrifugal force acting on the chuck jaws.

One of the disadvantages of these chucks is the tendency to increase the gripping pressure as the turning center slows down, which may damage the workpiece. An alternative method is to use elements of the chuck to lock the chuck jaws mechanically in their original position.

Counter-centrifugal chucks come in a variety of sizes from 8 to 18 in. (200 to 450 mm) in diameter and can operate at spindle speeds of 5500 r/min for an 8 in. (200 mm) diameter chuck and 3500 r/min for a 12 in. (300 mm) diameter chuck. The positioning repeatability of these chucks is .001 in. (0.02 mm).

Collet chucks

The collet chuck, Fig. 8-16-13, is ideal for holding finished square, hexagonal, and round bar stock. The collet assembly consists of a drawtube, a hollow cylinder with master collets, and collet pads. Master collets are available with three or four gripping fingers and are referred to as either three-split or four-split design. The four-split design has better gripping power, but is less accurate. The collet chucks are front-actuated, and the collet pads are sized for the diameter to be machined.

Fig. 8-16-13   A collet chuck provides precision holding for the workpiece. *(Hitachi Seiki Co. Ltd.)*

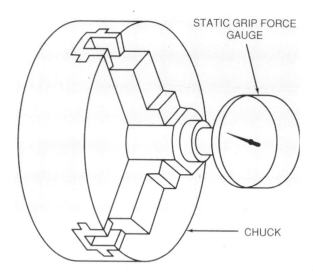

STATIC GRIP FORCE GAUGE

CHUCK

Fig. 8-16-14   Static gripping forces determines the force per jaw being exerted on the workpiece when the spindle is stopped. *(Cincinnati Machine, A UNOVA Co.)*

Chuck Jaw Clamping Force

Two devices can be used to measure the clamping force applied to the workpiece. One measures the static gripping force, Fig. 8-16-14, and one measures the dynamic gripping force. The static gripping force is the force per jaw exerted by the chuck on the workpiece when the spindle is stopped. The dynamic gripping force is the force per jaw exerted by the chuck on the workpiece when the spindle is running. Each individual chuck has a specific clamping force.

Chuck Jaw Pressure Limitations

When the chuck jaw clamping pressure is set, the pressure must not exceed the maximum pressure stamped on the warning plate. If a greater pressure is used, high stress forces are created within the chuck, resulting in possible damage to the chuck or to the machine, and possibly personal injury.

Typically, front-actuated chucks can be operated between 200 and 500 psi. Operating the chuck below 200 psi will cause insufficient clamping force on the part. The maximum chuck clamping pressure should always be used unless the pressure applied will damage the part.

Centrifugal Force and Speed limitations

Centrifugal force imposes speed limitations upon all types of chucks. Centrifugal force, which increases as the speed of rotation increases, tends to force the chuck jaws outward. This decreases the amount of the clamping force on the part. All chucks are affected by the high internal stresses caused by centrifugal force and therefore all have a recommended maximum rotation speed that should never be exceeded. Operating chucks at spindle speeds higher than recommended will result in higher internal stresses and the loss of clamping force on the part. High stresses can cause jaw or chuck component breakage that may release the clamping pressure of the jaws on the workpiece.

Centrifugal force increases as spindle speed increases and as jaws are made heavier, resulting in the jaws trying to move away from the centerline of rotation. Do not mount top jaws so that they extend beyond the diameter of the chuck. Also, reduce the spindle speed when using special top jaw tooling.

Changing Chuck Jaws

A quick-jaw changing system is used when it is necessary to frequently change the chuck jaws for different size workpieces, or for additional machining operations. This quick jaw-changing system, Fig. 8-16-15, can reduce the changing time from the usual 30 minutes to 1 1/2 minutes or less, which can pay for itself in a very short time. Machine downtime is less, and the productivity of the machine improves dramatically.

Chucking centers can also be equipped with an automatic chuck-changing system, Fig. 8-16-16. The more chuck-changing that is required, the more important the system becomes in reducing downtime. When the automated-jaw changing system is called up by the CNC program, it moves into position in front of the chuck. The jaws mounted in the chuck are removed and returned to their position in the magazine, and those required for the next operation are mounted into the chuck. Some automated systems change only one jaw at a time and other systems are capable of changing all jaws at once. Manually changing the chuck jaws could take the operator 20 to 30 minutes, but with new quick-change designs, the jaws or the inserts can be changed in 1 minute or less.

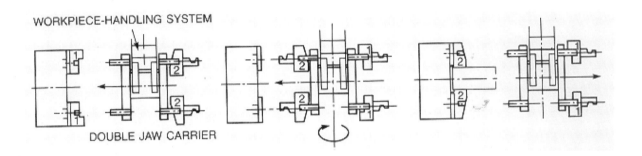

Fig. 8-16-15 A quick jaw-changing system can reduce the usual time of changing jaws from 30 min. to 1 1/2 min. or less. *(Rohm Products of America)*

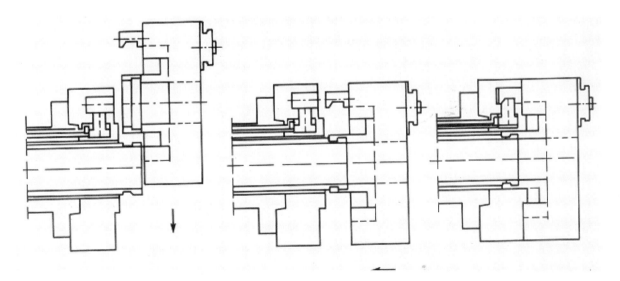

Fig. 8-16-16 An automatic jaw or chuck changing system reduces the amount of machine downtime and increases productivity. *(Forkardt Inc.)*

KNOWLEDGE REVIEW

Cutting Tools

1. What are the advantages of using indexable inserts?
2. Name three advantages of using wear-resistant tool coatings.
3. How do thin coatings affect the machining operation?
4. What is the general application for:
   (a) Titanium Nitride (TiN)?
   (b) Chromium Nitride (CrN)?
   (c) Tungsten Carbide/Carbon (WC/C)?
5. Name two basic types of carbide turning tools.
6. What type of carbide provides the best crater resistance?
7. List two main considerations when selecting a cutting tool insert.
8. What are six factors that affect the cutting tool life?
9. Name three advantages of using ceramic inserts.
10. What are cermet inserts?
11. List two main applications of cermet inserts.
12. What types of materials should be machined with CBN inserts?

Tooling Systems

13. What types of materials should be machined with diamond inserts?
14. Name the main parts of a typical tooling system.
15. For what purpose is preset tooling used?
16. How do manufacturers establish qualified tooling?

Tool Monitoring

17. What is the most common method of detecting tool wear?
18. What does in-process gaging monitor?

Speeds and Feeds

19. Name four factors that affect cutting speed.
20. List two problems that result from:
    (a) too slow a feed rate (b) too fast a feed rate.

Workholding Devices

21. Name the most common workholding device used on a turning center.
22. List three different types of chucks used on chucking centers.
23. Name two advantages of self-centering chucks.
24. List the advantages and disadvantages of counter-centrifugal chucks.
25. What is the repeatability of the counter-centrifugal chuck?
26. Name the type of workpieces best suited for machining with collet chucks.
27. Briefly define: (a) static gripping force (b) dynamic gripping force
28. Explain why correct chuck clamping pressure is important?
29. Why is it important not to operate chucks at speeds higher than their rated maximum speed?

Unit 17

CNC Bench-Top Lathe

CNC bench-top lathes are a relatively inexpensive way to teach and demonstrate the principles of CNC turning. These lathes are easy to program yet still have most of the options available on standard-size turning centers. The machines can be fitted with different types of controllers, depending on the machine tool manufacturer. The bench-top CNC lathe, Fig. 8-17-1, shown and described in this unit has a Fanuc compatible controller that is the most widely used controller in the world.

MACHINE FUNCTION CODES

Most of the preparatory functions (G codes) and the miscellaneous functions (M codes) used on industrial size turning centers can be used on the CNC bench-top lathe, with a few exceptions. The main difference is that on a standard CNC turning center the cutting tools are mounted to the rear of the workpiece, and on some bench-top lathes the cutting tools are mounted in front of the workpiece.

When programming a radius in a clockwise direction (CW) for a rear-mounted cutting tool lathe the standard G02 command is used; when programming the same radius with a front- mounted cutting tool a G03 command has to be used to produce the same radius, Fig. 8-17-2.

OBJECTIVES

After completing this unit, you should be able to:

1. Describe and state the operations that can be performed on a CNC lathe.
2. Use the proper G codes and M codes to write a CNC lathe program.
3. Write simple lathe programs for the CNC bench-top lathe.

KEY TERMS

G codes modal	non-modal
M codes	program format
modal	word address

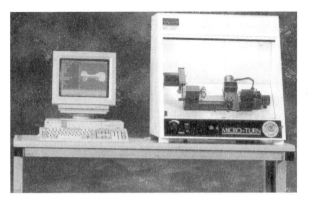

Fig. 8-17-1 CNC Bench-Top lathes are excellent teaching tools that are affordable and easy to program. *(Denford, Inc.)*

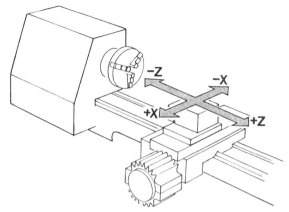

Fig. 8-17-3 Two main axes (X and Z) on a CNC lathe or turning center. *(Emco Maier, Corp.)*

MACHINE AXIS

The machine axis for the bench-top lathe is based on the same coordinate system as other CNC machines. Lathes are usually programmed in two axes; the **X** axis moves the cutting tool or carriage towards (**X-**) or away (**X+**) from the spindle centerline. The **Z** axis moves the carriage or cutting tool along the workpiece (**Z-**) toward the chuck or headstock and (**Z+**) away from the chuck, Fig. 8-17-3.

MODAL AND NON-MODAL CODES

All G codes and M codes fall into two categories: modal and non-modal.

Modal: Modal codes are codes that stay in effect until they are cancelled by a code such as G80

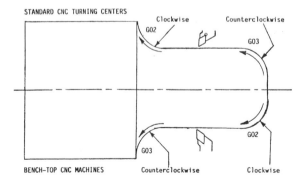

Fig. 8-17-2 Comparison of circular interpolation G02 and G03 codes on CNC conventional and bench-top turning centers. *(Kelmar Associates)*

which cancels the G81 (drilling cycle), or they are replaced with another code of the same group or family, such as G00, G01, G02, G03.

Non-modal: Non-modal codes are codes that stay in effect only in the block they are programmed in, such as G04 (non-modal dwell time).

PREPARATORY COMMANDS G CODES

The program address G plus two digits (numbers) is a preparatory command used to prepare or preset the control system to accept the instructions following it. The address G01 sets the control for linear interpolation; G20 presets the control for inch measurement. Many G codes are used on a variety of machine tools and each tells the control (machine) to perform a specific function. Even though most G and M codes apply to most machines, there are variations between manufacturers and it is wise to check the CNC machine specifications that apply to each machine being used.

G codes (bench-top lathes)

CODE	GROUP	FUNCTION
G00	1	Positioning (Rapid Traverse)
G01	1	Linear Interpolation (Feed)
G02	1	Circular Interpolation (CW)
G03	1	Circular Interpolation (CCW)

G04	0	Dwell
G20	6	Inch Data Input
G21	6	Metric Data Input
G28	0	Reference Point Return
G40	7	Tool-Nose Radius Compensation Cancel
G41	7	Tool-Nose Radius Compensation Left
G42	7	Tool-Nose Radius Compensation Right
G50	0	Work Coordinate Change/ Max. Spindle, Speed setting
G70	0	Finishing Cycle
G71	0	Stock Removal in Turning - X
G72	0	Stock Removal in Facing - Z
G73	0	Pattern Repeating
G74	0	Peck Drilling in Z Axis
G75	0	Grooving in X Axis
G76	0	Multiple Thread Cutting Cycle
G81	1	Deep hole drilling (No FANUC)
G90	1	Diameter Cutting Cycle A (Outer/Inner)
G92	1	Thread Cutting Cycle
G94	1	Cutting Cycle B (End Face Cycle)
G96	2	Assessed Surface Speed Control
G97	2	Assessed Surface Speed Control Cancel
G98	11	Feed Per Minute
G99	11	Feed Per Revolution

The G codes are organized in groups with the group number following the G code. G codes belonging to the same number group cancel each other, for example, G20 cancels G21, and G21 cancels G20. Group 0 (zero) is a group of non-modal G codes.

MISCELLANEOUS FUNCTIONS M CODES

The program address M followed by two digits (numbers) or three numbers on some machines identifies a miscellaneous function. Some M functions relate to the operation of the machine, and some relate to the operation of the program. Although the term modal is not used with M codes, some codes will remain in effect when used e.g. M03 (spindle ON), M09 (coolant pump OFF). Others are only active in the block in which they are used e.g. M00 (program stop). It is important for the programmer to know when the M code will take effect, whether it is the block programmed or in the next block. Always consult the programming manual supplied with the machine tool and the controller for the exact definition and use of M codes.

M codes (Bench-Top Lathe)

CODE	FUNCTION
M00	Program stop
M01	Optional stop
M02	End of program (no return to the top of program)
M03	Spindle forward (CW)
M04	Spindle reverse (CCW)
M05	Spindle stop
M08	Coolant ON
M09	Coolant OFF
M10	Chuck open
M11	Chuck close
M13	Spindle forward and coolant ON
M14	Spindle reverse and coolant ON
M25	Tailstock quill extend
M26	Tailstock quill retract
M30	Program stop, return to top of program
M38	Door open
M39	Door close
M40	Parts catcher extend
M41	Parts catcher retract
M98	Sub program call.
M99	Sub program end

Various auxiliary codes exist that can control different devices such as a robot and other accessories; consult the machine manual for these.

WORD ADDRESS FORMAT

The word address format is based on a combination of letters and one or more numbers. A word consists of an address, which is one of twenty-six (26) letters from A to Z followed by the numerical data of the word. The numerical data following the address could represent a sequence number N015), a tool offset (H01) or a feed rate (F10.0). Each program block can contain one or more words. The block number must be the first word, usually followed by a G code(s), then the primary axis X and Z.

Some controls require use of a specific format or sequence for words in a block. Other controls will accept them in random sequence.

The word address format for the CNC benchtop lathe used in this unit is:

N   block number.
G   G code (preparatory function).
X   Absolute distance traveled by tool in X axis.
U   Incremental distance traveled by tool in X axis.
Z   Absolute distance traveled by tool in Z axis.
W   Incremental distance traveled by tool in Z axis.
F   Feed rate.
M   M code (miscellaneous function).
S   Spindle speed.
T   Tooling management.

Each block, or program line, contains addresses, which appear in this sequence N, G, X (or U), Z (or W), F, M, S, T. This sequence should be maintained throughout every block in the program, although not all blocks may contain all the addresses.

PROGRAMMING PROCEDURE

Before the programmer/operator of the CNC lathe starts machining, it is his/her responsibility to see that the machine and the cutting tools are setup correctly. Regardless of how good the CNC program, or how accurate the machine is, it cannot produce a precision part if the wrong tools are used or the machine is not setup correctly.

Some steps to follow in setting up the lathe are to check:

1. that the program has been loaded into the MCU correctly.
2. that the cutting tools are the correct ones for the program.
3. that the correct insert is mounted in the cutting toolholder.
4. that the X and Z reference points are the same as the programmer's.
5. that the X and Z zero references are registered in the controller.
6. that the tool offsets are correct for the tools used.
7. that the type and size of material is correct?

NOTE: It is the CNC operator's responsibility to be sure that everything has been checked so that the finished part is accurate and meets the specifications and quality required.

The program O2071 will identify each code and movement made in the program to produce the finished part. This check will help the learner to understand what is happening as a result of the various codes in each program block.

PROGRAMMING NOTES:

1. The material used for the project is 1.000 in. diameter aluminum bar stock 3.000 in. long.
2. Spindle speed is 2000 r/min.
3. Carbide insert cutting tool.
4. Feed rate (roughing) .008 in. per revolution.
5. Feed rate (finishing) .005 in. per revolution.

TURNING PROGRAMMING – (Fanuc Compatible Controller)

Programming Sequence

O2071		Program number
N010	**G20 G40**	
	G20	Inch data input

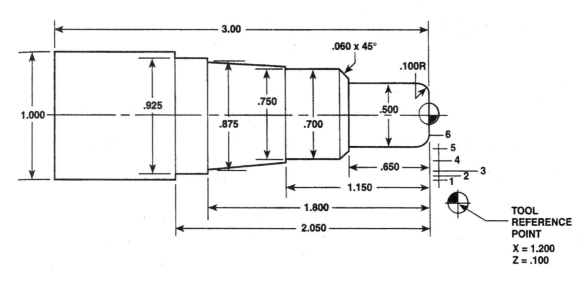

Fig. 8-17-4 A sample CNC turning project that uses a variety of basic G and M programming codes.
(Kelmar Associates)

G40   Cancels tool radius compensation.

N020   **G95 G96 S200 M03**

G95   Feed rate per revolution.

G96   Constant feed rate.

S200   Spindle speed 200 ft/min.

M03   Spindle **ON** clockwise.

N030 **T0202**   Tool number and offset number.

N040   **G00 X1.200 Z.100**

G00   Rapid traverse mode.

X & Z   Tool reference or change point.

X1.200 Tool point .100 away from outside diameter.

Z.100   Tool point .100 to right of end of work.

Rough Turning Cycle

N050   **G73 U.05 R.05**

G73   Rough turning cycle.

U.05   Depth of the roughing cut (per side)

R.05   Retract amount from each cut

N060   **G73 P35 Q95 U.04 W.005 F.008**

P35   Start block of rough contour cycle.

Q95   End block of rough contour cycle.

U.04   Stock allowance on diameter

W.005   Shoulder allowance for finish cut.

F.008   Feed rate .008 per revolution.

N070   **G00 X.300 Z.050**

G00   Rapid traverse mode.

X.300   Tool point at .300 dia. for .100 radius start.

Z.050   Tool point .050 away from end of part.

N080   **G01 Z0**

G01   Linear interpolation (feed).

Z0   Tool point touching end of work.

N090   **G03 X.500 Z-.100 R.100**

G03   Circular interpolation (CCW).

X.500   Largest diameter of radius.

Z-.100 End of radius on .500 diameter.

R.100   Size of radius.

N100   **G01 Z-.650**

G01   Linear interpolation.

Z-.650 Machines .500 diameter to .650 length.

N110   **X.580**

X.580   Tool moves to small diameter of .060 x 45° bevel.

N120	X.700 Z-.710
	X.700 Large diameter of bevel.
	Z-.710 End distance of bevel.
N130	**Z-1.150**
	Z-1.150 .700 diameter cut to 1.150 length.
N140	**X.750**
	X.750 Cutting tool feeds to .750 (small end of taper).
N150	**X.875 Z-1.800 (cutting taper)**
	X.875 Large end of taper.
	Z-1.800 Length taper is cut.
N160	**X.925**
	X.925 Tool feeds out (faces) to .925 diameter.
N170	**Z-2.050**
	Z-2.050 .925 diameter cut to 2.050 length.
N180	**X1.050**
	X1.050 Tool feeds to .050 past part diameter.
N190	**G00 X 1.200 Z. 100 (tool back to tool reference point)**
	G00 Rapid traverse mode.
	X1.200 Z.100 (reference point position)

Finish Turning

N200   G70 P35 Q95 F.005

	G70 Finish turn cycle.
	F.005 Feed rate .005 per revolution.
N210	**G00 X2.000.Z.500**
	G00 Rapid traverse mode.
	F.005 Feed rate .005 per revolution.
N220	**M30**
	M30 End of program
%	Rewind code.

SUMMARY

- Bench-top lathes provide an easy, inexpensive way to teach the principles of CNC lathe programming.
- These lathes are quick and easy to program and have most of the options that are available on a standard-size machine.
- The machine controllers are compatible with those used in industry on full-size machine tools.
- G-codes and M-codes are interchangeable between the bench-top lathe and a standard-size turning center, except for a few variations.
- The size of the machine makes it less intimidating and more user friendly for a first time user of CNC machines.
- Bench-top lathes are more affordable for educational institutions than industrial-size machine tools.

KNOWLEDGE REVIEW

1. List three advantages of a bench-top lathe.
2. Define the following preparatory G codes: G01, G02, G20, G40, G70, G96.
3. What are the following miscellaneous code functions: M00, M03, M05, M08, M10, M42, M98?
4. Define the motion of the **X** axis.
5. Define the motion of the **Z** axis.
6. Explain the terms: modal and non-modal.
7. List five items that should be checked by the operator before starting to machine a part.
8. In program O2071, what is the programmed spindle speed and feed rate for finish turning?

Unit 18

CNC Chucking and Turning Centers

Numerically controlled (NC) lathes had a very slow start in manufacturing. Studies during the 1960's indicated that 40% of all metal cutting operations were performed on lathes, yet NC lathes accounted for only 7.4% percent of the turning products produced.

The early NC lathes were standard engine lathes that had been retrofitted with control systems and other components to convert a conventional lathe into a numerical control lathe. These lathes were a big improvement over the conventional and tracer lathes being used at that time. They were capable of making contour cuts by controlling the coordinated motion of the cross-slide and the carriage. Thread cutting was made possible by the automatic synchronization of the spindle revolutions per minute (r/min) and the travel of the carriage.

With the introduction of the computer, the NC lathe became computerized numerically controlled (CNC) chucking and turning centers that were capable of better precision and higher production rates. These machine tools were equipped with color graphics display screens for programming, in-process gaging, tool changers, machining monitors, rotary tooling, and automatic loading and unloading devices. This allowed the machine to run virtually unattended with a minimum of downtime, making CNC chucking and turning centers exceptionally versatile machine tools.

OBJECTIVES

After completing this unit you should be able to:

1. Describe the purpose and functions of chucking and turning centers.
2. List the applications of CNC for chucking and turning centers.
3. Identify the types of machining operations for which each machine is designed.

KEY TERMS

function codes	indexable turrets
material handling	tailstocks
tool orientation	tool-nose radius compensation

TYPES OF TURNING CENTERS

Computer Numerical Control (CNC) lathes are classified in different ways, such as: the type of machine (horizontal or vertical), the location of the tooling (front or back), and the number of axes (two, three, four, or six). The horizontal type is the most common type used in manufacturing and is available with either flat- or slant-bed options.

CNC Chucking Center

The CNC chucking center, Fig. 8-18-1, is designed to machine most work that is held in a chuck. These machines are manufactured in a wide variety of sizes, from bench-top machines, Fig. 8-18-2, to machines with chuck sizes ranging from 8 to 36 in. (200 to 900 mm) in diameter, spindle drive motors from 5 to 75 horsepower, and spindle speeds up to 6000 r/min.

CNC Turning Center

CNC turning centers, Fig. 8-18-3, similar to chucking centers, are designed for machining shaft-type workpieces that are supported by a chuck and a heavy-duty tailstock center.

Fig. 8-18-1   CNC chucking center holds the workpiece in a chuck or fixture. *(Hardinge, Inc.)*

Fig. 8-18-2   CNC bench-top lathe is used primarily for teaching purposes. *(Denford, Inc.)*

Two-axis turning center

The two-axis CNC turning center is the most commonly used turning centers. The workholding device is usually a chuck mounted on the left side of the machine spindle as viewed by the operator. The slant-bed lathe with rear-mounted tooling is the most popular design for general-purpose machining. The cutting tools, Fig. 8-18-4, are held in an indexable turret designed to hold four, six, eight, ten, twelve, or more tools.

Three-axis turning center

The three-axis CNC lathe is basically a two-axis lathe with live tooling as the additional axis. This axis, that has its own power source, is fully programmable and is designated as the **C** axis. It is used for drilling bolt-hole circles, slot milling, milling flat surfaces and helical slots. The **C** axis can perform some drilling and milling operations on the part while still in the turning center, eliminating the need to move the part to another machine.

Four-axis turning center

The four axis-turning centers use two turrets operating independently on separate slides, machining the workpiece simultaneously. While the upper turret is machining the inside diameter,

Fig. 8-18-3 CNC turning centers are used mainly for machining shaft-type workpieces.
(Cincinnati Machine, A UNOVA Co.)

the lower turret may be machining the outside diameter, Fig. 8-18-5. The turrets balance the cutting forces applied to the work so that extremely heavy cuts can be taken on a workpiece when it is supported by the tailstock. However, if the

Fig. 8-18-4 The two-axis turning center is most commonly used in industry. *(Hardinge, Inc.)*

Fig. 8-18-5 Turning and boring operations may be done simultaneously. *(Cincinnati Machine, A UNOVA Co.)*

Fig. 8-18-6   Two turrets simultaneously machining the outside diameter of a workpiece. *(Cincinnati Machine, A UNOVA Co.)*

Fig. 8-18-7   Two turrets performing turning and facing operations. *(Cincinnati Machine, A UNOVA Co.)*

workpiece requires mainly internal operations, both turrets can work on the inside of the workpiece at the same time. This type of operation is suitable for large-diameter parts that require boring, chamfering, threading, internal radii, or retaining grooves.

For parts with mostly outside diameter operations, the upper turret can be equipped with turning tools so that both turrets can machine the outside diameter on a shaft at the same time, Fig. 8-18-6.

When longer parts must be machined, the right hand end of the shaft may be supported with a center mounted in the upper turret while the lower turret performs the external machining operations. Other operations that may be performed simultaneously are turning and facing, Fig. 8-18-7, and internal and external threading. When the turning center is equipped with a steadyrest, operations such as facing and threading may be performed on the end of a shaft, Fig. 8-18-8.

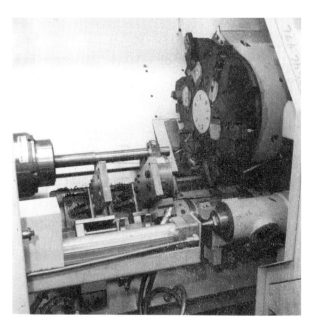

Fig. 8-18-8   A steadyrest supporting a workpiece so that machining may be performed on the end of the shaft. *(Cincinnati Machine, A UNOVA Co.)*

Fig. 8-18-9 The slant bed design provides for easy operator access and chip removal.
(*Cincinnati Machine, A UNOVA Co.*)

Six-axis turning centers

Six axis lathes are specially designed lathes, with twin turrets and a set of three axes per turret. This design incorporates many tool stations, some of them power driven, which allows for back machining. Programming these lathes is similar to programming a three-axis lathe twice. The control system provides the synchronization when necessary. Small to medium size six-axis lathes are used in screw machine shops, and industries with similar applications.

CNC CHUCKING AND TURNING CENTER PARTS

The main parts of the CNC chucking and turning centers are the bed, headstock, cross-slide, carriage, turret, tailstock, servomotors, ball screws, hydraulic and lubrication systems, and the machine control unit (MCU).

Bed: The bed is usually made of high-quality cast iron, which is well suited to absorb the shocks created by heavy cuts. It is usually a slant bed design, Fig. 8-18-9, slanting from 30 to 45°, providing easy access for the operator for loading and unloading of the parts. The design also allows the chips and coolant to fall away from the cutting area to the

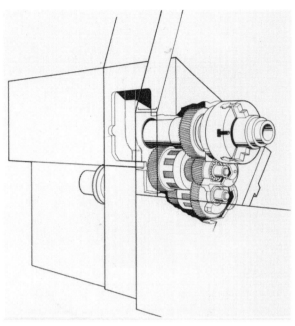

Fig. 8-18-10 The headstock transmits maximum horsepower and torque to the spindle.
(*Cincinnati Machine, A UNOVA Co.*)

bottom of the bed. Parallel surfaces are machined into the front of the cast bed, providing mounting tracks for the hardened bed ways.

Headstock: The headstock of the CNC chucking and turning centers, Fig. 8-18-10, transmit the maximum horsepower and torque from the motor to the spindle. These machines are available with a variety of motor sizes ranging from 5 to 75 horsepower and spindle speeds from 32 to 6000 revolutions per minute (r/min). The spindle speed is usually programmable in 1 r/min increments.

Tailstock: CNC chucking and turning centers can be equipped with a manual tailstock similar to a standard engine lathe, or a programmable swing-up tailstock. The tailstock travels on its own hardened and ground bearing ways, allowing the carriage to move past the tailstock when a short shaft is being held. The need to extend the quill of the tailstock to its maximum distance is also eliminated, thus maintaining greater rigidity of the part.

Fig. 8-18-11 A programmable tailstock can be moved manually or by a programmed command. *(Cincinnati Machine, A UNOVA Co.)*

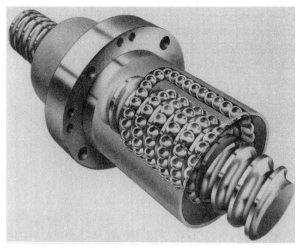

Fig. 8-18-13 Precision ball screws provide accurate slide positioning. *(Cincinnati Machine, A UNOVA Co.)*

Fig. 8-18-12 Turret holding both inner and outer turning tools. *(Cincinnati Machine, A UNOVA Co.)*

Fig. 8-18-14 The machine control unit allows the operator to program a part, check it for accuracy, and simulate the machining operation to check for errors. *(Traub GmbH Index Corp.)*

Programmable tailstock: The tailstock in Fig. 8-18-11 can be moved manually or by computer command. Positioning, clamping, and unclamping of the tailstock to the bearing ways is done by hydraulic pressure. The tailstock is protected against collision with the indexing tools by a contact sensor that immediately stops the indexing motion on contact.

Swing-up tailstock: This tailstock adds flexibility and versatility to either the chucking or the turning center. It can swing up to support workpieces for external machining, and then swing away to allow the machine to perform internal work such as deep-hole drilling and boring.

Turrets: The type, style, and number of turrets on CNC chucking or turning centers varies according to the size of the machine and individual manufacturer's specifications. The more common types used are the drum turret, disk turret, and the square multi-tool turret. The turret is constructed so that it can hold twelve or more inner and outer turning tools Fig. 8-18-12. The indexing mechanism for these tools is capable of bi-directional indexing and a rapid traverse rate of approximately 400 in./min (100 r/mm), which minimizes the non-cutting time.

Servo system: The servo system, Fig 8-18-13, consisting of servo drive motors, ball screws, and rotary resolvers, provides the fast, accurate movements and positioning of the **X** and **Z** axes slides. The rotary resolvers provide the system with unidirectional slide positioning repeatability of ±.0002 in. (0.005 mm). The rapid movement of the slides is approximately 950 in./min (24,000 mm/min), and the feed rates are programmable in .l00 in./min (2.54 mm./min) increments.

Machine control unit (MCU): The MCU, Fig. 8-18-14, allows the operator to program a part, edit a program, graphically display programs, store programs into memory, output programs to a computer or data line, perform comprehensive diagnostics, run a program manually or automatically, and perform many more functions and operations. These controls are designed and manufactured using the latest "state-of-the-art" technologies and features and vary accordingly to individual manufacturer's requirements and specifications.

Steadyrest/follower rest: When long, thin workpieces or shafts are machined, chatter usually occurs. The chatter or vibration can be minimized by means of a steadyrest or a follower rest, Fig. 8-18-15.

The steadyrest is mounted to the bed of the turning center and can be programmed to open or close automatically providing feed-through capabilities for the workpiece. The steadyrest uses constant hydraulic pressure applied to the support rollers enabling it to adjust to variations in workpiece diameter. The support provided by

Fig. 8-18-15 **Two follower rests being used to support a long thin shaft.** *(Cincinnati Machine, A UNOVA Co.)*

the steadyrest reduces chatter in the workpiece and allows the machine to operate at higher and more efficient speeds and feeds. When a four-axis chucking or turning center is used, a follower rest can be mounted in the lower turret. The turret can be programmed to move at the desired rate of travel, providing constant support to the workpiece while the cutting tool mounted in the upper turret does the machining.

MATERIAL-HANDLING SYSTEMS

Several options can be added to the CNC chucking and turning centers to enhance their performance and productivity. Some of these options are the bar feeder, parts catcher, parts loader/unloader, chip conveyor, and robot loader.

- The **bar feeder** is capable of handling 6 ft (2 m) and 12 ft (4 m) bar lengths, eliminating the loading of individual part blanks.
- The **parts catcher**, Fig. 8-18-16, complements the bar feeder and deposits the machined parts outside the machine. When the bar feeder and parts catcher are used, the loading and unloading time is reduced.
- The **parts loader/unloader**, Fig. 8-18-17, allows

Fig. 8-18-16 Parts catcher delivers the finished parts outside the machine. *(Cincinnati Machine, A UNOVA Co.)*

Fig. 8-18-18 A robot loader being used for programmed work handling for a turning center. *(Cincinnati Machine, A UNOVA Co.)*

individual part blanks of .75 to 2.00 in. (20 to 50 mm) in diameter and 1.00 to 2.00 in. (25 to 50 mm) in length to be loaded and unloaded in approximately six seconds.

- The **chip conveyor** picks up all the chips from the bed of the machine. The chips produced by the cutting cycle fall freely onto the chip conveyor track because of the slant bed design of the machine. The chips are then transported by the conveyor system out of the bottom of the machine into containers for storage and recycling.
- The **robot loader**, Fig. 8-18-18, communicates with the MCU and performs operations such as loading and unloading parts, storing and retrieving parts from pallets, transporting parts to gauging stations, and changing chuck jaws. Dedicated robot loaders represent the major trend in automated workhandling for turning centers.

Fig. 8-18-17 A part loader/unloader places and removes individual parts from the machine. *(Cincinnati Machine, A UNOVA Co.)*

MACHINE AXES

Horizontal CNC lathes are usually designed with two programmable axes, but can also have three, four or six axes. The two primary axes of the CNC lathe are the **X** and **Z** axes. The **X** axis is the cross travel of the cutting tool, toward and away from the spindle centerline. The **Z** axis is the longitudinal motion of the cutting tool toward and away from the headstock or chuck.

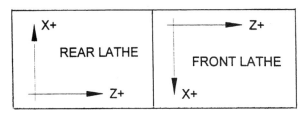

Fig. 8-18-19 CNC lathe axes for a front and rear lathe/turning center. *(Peter Smid)*

The **C** axis (live tooling) is usually the third axis and is designed for drilling and milling operations, Fig. 8-18-19. It is an optional feature on typical two-axis lathes and its operations are fully programmable.

FUNCTION CODES

Many preparatory and miscellaneous function codes are the same for turning centers, machining centers, and wire-cut electrical discharge machines (EDMs). However, because of the nature of work produced on each type of machine, certain differences exist in the coding systems to suit each machine. It is very important that the programmer recognize that differences do exist in order to properly program each machine. The use of an incorrect code might result in scrapped work, damage to the machine or cutting tool, or no response from the MCU.

Preparatory Codes for Turning Centers

The G-code is a preparatory command used to prepare or preset the control system to a certain condition or mode of operation. The G address prepares the control to accept the instruction following the code.

NOTE: Most Fanuc lathes in North America use the Type "A" G-codes. Only one type is set per controller during installation.

G-codes used for Turning and Chucking Centers are listed as follows:

G00	Rapid positioning
G01	Linear interpolation
G02	Circular interpolation clockwise (CW)
G03	Circular interpolation counterclockwise (CCW)
G04	Dwell (in a separate block)
G10	Programmable data input (data setting)
G11	Data setting mode (cancel)
G20	Inch data input
G21	Metric data input
G27	Machine zero return check
G28	Machine zero return (reference point 1)
G29	Return from machine zero
G30	Machine zero return (reference point 2)
G31	Skip function
G32	Thread cutting - constant lead
G34	Thread cutting - variable lead
G35	Circular threading (CW)
G36	Circular threading (CCW)
G40	Tool tip radius compensation cancel
G41	Tool tip radius compensation left
G42	Tool tip radius compensation right
G50	Tool position register (max. r/min preset)
G52	Local coordinates preset
G53	Machine coordinate system setting
G54-G59	Work coordinates system selection
G61	Exact stop mode
G62	Automatic corner override mode
G64	Cutting mode
G65	Custom macro call
G70	Profile finish cycle
G71	Profile rough cutting cycle (horizontal)
G72	Profile rough cutting cycle (vertical)
G73	Closed loop cutting cycle
G74	Drilling cycle
G75	Grooving cycle
G76	Thread cutting cycle
G90	Cutting cycle A (group A)
G90	Absolute Positioning (group B)
G91	Incremental Positioning

G92   Tool position register (group B)
G94   Cutting cycle B (group A)
G94   Feed rate per minute (group B)
G96   Constant surface speed (CSS)
G97   Direct r/min input (cancel CSS)
G98   Feed rate per minute (group A)
G99   Feed rate per revolution

Miscellaneous Functions for Turning Centers

The M-code is a miscellaneous function code used to activate certain machine tool operations or control the program flow. Most M-codes function like switches turning an activity **ON** or **OFF**, moving a device **IN** or **OUT**, forward or backward. M-codes used for Turning and Chucking Centers are as follows:

M00   Compulsory program stop
M01   Optional program stop
M02   End of program (reset, no rewind)
M03   Spindle rotation normal
M04   Spindle rotation reverse
M05   Spindle stop
M07   Coolant mist **ON**
M08   Coolant pump motor **ON**
M09   Coolant pump motor **OFF**
M10   Chuck unclamping (**OPEN**)
M11   Chuck clamping (**CLOSED**)
M12   Tailstock spindle (**IN**)
M13   Tailstock spindle (**OUT**)
M17   Turret rotation forward
M18   Turret rotation reverse
M19   Spindle orientation (optional)
M21   Tailstock forward
M22   Tailstock backward
M23   Chamfering **ON**
M24   Chamfering **OFF**
M30   Program end (reset and rewind)
M31   Chuck bypass **ON**
M32   Chuck bypass **OFF**
M41   Spindle speed (low range)
M42   Spindle speed (high range)
M48   Feed rate override cancel function (**OFF**)
M49   Feed rate override cancel function (**ON**)
M73   Parts catcher **OUT**
M74   Parts catcher **IN**
M98   Call subprogram
M99   End subprogram

TOOL NOSE RADIUS COMPENSATION

The radius compensation commands used on the CNC lathe are the same preparatory G-codes that are used for programming CNC mills, Fig. 8-18-20.

- **G41** – compensation of the tool-nose radius to the left of the contouring direction.
- **G42** - compensation of the tool-nose radius to the right of the contouring direction.
- **G40** – cancels the tool-nose radius compensation either in the left or right direction.

These codes allow the MCU to accurately produce arcs and tapers on the workpiece by automatically accounting for the size of the radius on the tool nose. Without tool-nose radius compensation, the workpiece profiles cut by the tool nose would be subject to undercutting and overcutting.

Tool Nose

The tool nose on turning tools is the corner of the cutting tool or insert where the cutting edges blend together. This is the point that is pro-

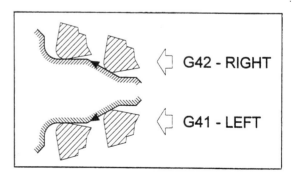

Fig. 8-18-20 **Tool nose radius compensation for lathe applications.** *(Peter Smid)*

grammed, and is called the command point or the imaginary point. It is referred to as the imaginary point because in reality the cutting-tool insert does not have a sharp point, Fig. 8-18-21. There is usually a radius on the insert to provide strength and longer life during the cutting operation. Cutting tools used for turning have a small radius usually:

.015 in. (0.4 mm)
.030 in. (0.8 mm)
.047 in. (1.2 mm)

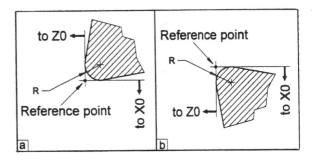

Fig. 8-18-21 Tool reference point. (A) a typical external tool, (B) a typical internal tool. *(Peter Smid)*

Tool Orientation

Unlike an end mill that is represented by a circle with the center always an equal distance from the cutter radius (cutting edges). CNC lathe cutting tools are different in that they have a radius and also have cutting edges that are not part of the radius. Although the center of the radius is always an equal distance from the part contour the cutting edges are oriented differently e.g. for turning and for boring, even if the tool-nose radius is the same size.

The setting of the reference point of the tool is at a different position on the radius for turning than it is for boring. If the center of the tool radius is located from the reference point, a vector of direction or tool tip orientation will be established. The same typical tools used previously show the direction of vectors and the tool tip orientation.

The machine control unit must be informed about the difference between these two tools otherwise it does not know the location of the tool-tip orientation. Fanuc and similar controls assign a fixed number for each possible tool-tip position, Fig. 8-18-22. This number has to be entered into the offset screen at the control, under the T or TIP heading. The tool-tip orientation number and the value or size of the tool radius must be entered in the machine-control system.

NOTE: If the tool-tip number is 0 (zero) or 9, the control will compensate to the center. This method is usually common to programs prepared on CAD/CAM systems.

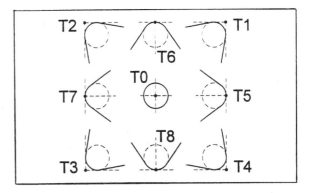

Fig. 8-18-22 Arbitrary tool tip numbers for tool nose radius compensation. The dot is the point of reference. *(Peter Smid)*

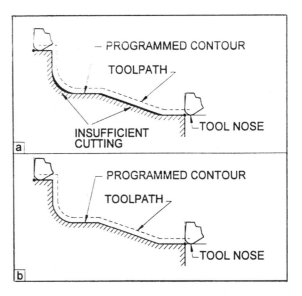

Fig. 8-18-23 Effect of tool nose radius compensation. (A) compensation not used, (B) compensation used. *(Peter Smid)*

Effects Of Radius Compensation

It is a mistake to think that because the radius on the insert or cutting tool is too small to need tool nose radius compensation. Fig. 8-18-23 shows a part that has been machined with and without the tool-nose radius compensation. All radii chamfers and tapers will be incorrectly machined without tool-nose radius compensation and could result in scrapped parts.

PART ZERO – TURNING CENTERS

CNC turning centers usually have only two axes the **X** and the **Z**, which makes the selection of the part zero relatively easy. Because the turning center produces cylindrical parts the **X** axis is always the spindle centerline. The **Z** axis has three popular locations the chuck face, the chuck jaw face, or the part face. The most popular location for the part zero for turning centers is the end face of the workpiece.

PROGRAMMING PROCEDURES

Turning centers are manufactured by a number of different manufacturers, and each of these machine tool builders can use a different type of controller. With each machine tool supplied there will be a complete programming manual containing the information required to operate and program the machine tool. Before the operator/programmer starts machining it is his/her responsibility to see that the proper programming and setup information has been entered into the MCU. Due to the number of different machines and controllers available it is likely that each machine will vary slightly in the way it is programmed and set up.

SAMPLE MACHINING PROGRAM

Turning centers manufactured by machine tool builders will vary; some mount the cross-slide on the slant bed, and others mount the cross-slide on a vertical support. Regardless of the construction of the machine, they all operate on the **X** axis (cross-slide movement) and **Z** axis (longitudinal movement). The **X** positive (**X+**) moves the cut-

ting tool toward the centerline of the spindle, and the **X** negative (**X–**) moves the cutting tool away from the spindle centerline. The **Z** positive (**Z+**) moves the saddle or cutting tool away from the headstock, while a **Z** negative (**Z–**) moves the saddle or cutting tool toward the headstock. Both incremental and absolute programming can be used on most turning centers.

The sample part shown in Fig. 8-18-24A, will be used to program some common machining operations performed on a chucking or turning center. Two passes will be made over the workpiece: one for rough turning and the other for finish turning operations.

The machining program

O3001		Program number
N010	**G20**	
	N010	Sequence number
	G20	Inch input data

Roughing cut

N020	**G50 X-2.0 Z2.5 S1000**	
	G50	Absolute coordinate preset, the reference point for start point of individual cutting tools.
	S1000	Maximum spindle speed 1000 r/min.
N030	**G00 T0101 M41**	
	G00	Rapid positioning
	T01	Tool number 01 activate offset number 01.
	M41	Spindle speed low range.
N040	**G96 S100 M03**	
	G96	Constant surface speed.
	S100	Speed 100 sf/min.
	M03	Spindle rotation clockwise (CW)
N050	**G00 X-2.02 Z.1 M08**	
	G00	Rapid move **a-b**.
	M08	Coolant pump motor **ON**.
N060	**G01 Z.01 F.120**	
	G01	Linear move **b-c**.
	F.120	Feed rate 120 in./r.
N070	**Z2.49 F.012**	
		Linear move **c-d**.

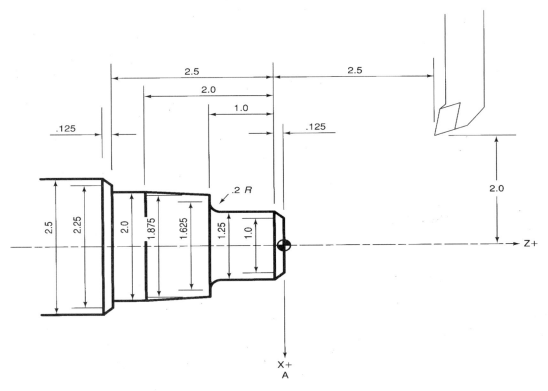

Fig. 8-18-24A   A sample part that can be machined on a turning center. *(Kelmar Associates)*

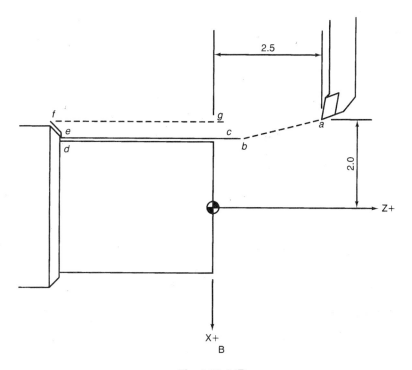

Fig. 8-18-24B

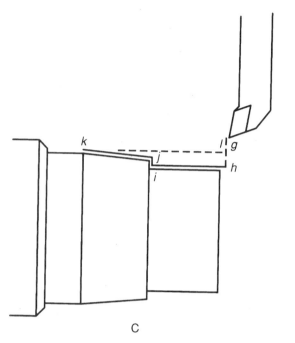

Fig. 8-18-24C

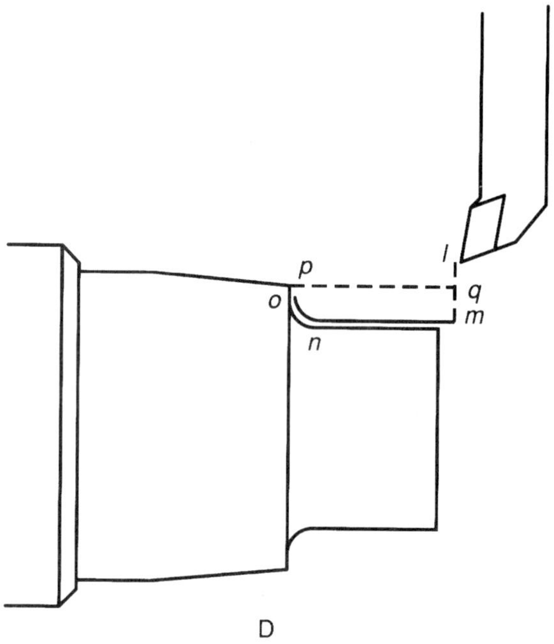

Fig. 8-18-24D

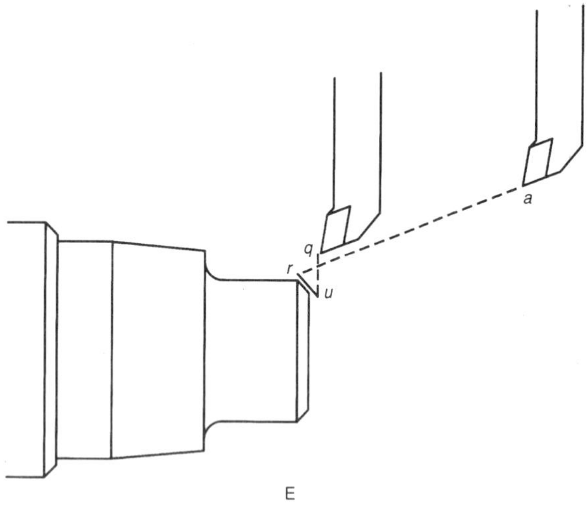

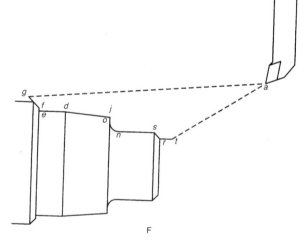

Fig. 8-18-24E

Fig. 8-18-24F

N080	**X-2.27**	
	Linear move **d-e**.	
N090	**X-2.52 Z-2.625**	
	Linear move **e-f** (chamfer)	
N100	**G00 Z.01**	
	G00Rapid move **f-g**.	
N110	**X-1.645**	

	Rapid move **g-h**	
N120	**G01 Z.99**	
	G01 Linear move **h-i**	
N130	**X-1.895**	
	Linear move **i-j**	
N140	**X-2.02 Z-2.0**	
	Linear move **j-k** (taper)	

N150	G00 Z.01
	G00 Rapid move **k-i**
N160	X-1.27
	Rapid move **l-m**
N170	G01 Z.79
	G01 Linear move **m-n**.
N180	G03 X-1.645 Z.99 R.2
	G03 Circular interpolation clockwise (CW) **n-o**.
	R.2 .200 radius.
N190	G01 X-1.895
	G01 Linear move **c-p**.
N200	G00 Z.01
	G00 Rapid move **p-g**.
N210	X-1.02
	Rapid move **q-m**.
N220	G01 X-1.27 Z-.135
	G01 Linear move **m-r** (chamfer)
N230	G00 X-4.0 Z2.5 M05
	G00 Rapid move **r-a**.
	M05 Spindle stop.
N240	T0100
	T0100 Tool number 01 activate offset number 00 (0 cancel tool offset)

Finishing cut

N250	G50 X-2.0 Z2.5 S2000
	G50 Absolute coordinate preset for cutting tools.
	S2000 Spindle speed 2000 r/min.
N260	T0202 M42
	T02 Tool number 02 activate offset number 02.
	M42 Spindle speed high range.
N270	G96 S150 M03
	G96 Constant surface speed.
	S150 Speed 150 sf/min.
	M03 Spindle on clockwise (CW).
N280	G00 X-1.0 Z.2
	G00 Rapid move to start of finish cut **a-t**.
N290	G01 Z.1 F.040
	G01 Linear move **t-r**.
	F.040 Feed rate .040 in./r.

N300	X-1.25 Z-.125 F.006
	Linear move **r-s** (chamfer)
	F.006 Feed rate.
N310	Z-.8
	Linear move **s-n**.
N320	G03 X-1.625 Z-1.0 R.2
	G03 Circular interpolation clockwise (CW) **n-o**
	R.2 .200 radius
N330	G01 X-1.875
	G01 Linear move **o-j**
N340	X2.0 Z-2.0
N340	X2.0 Z-2.0
	Linear move **j-d** (taper)
N350	Z-2.5
	Linear move **d-e**
N360	X-2.25
	Linear move **e-f**
N370	X-2.5 Z-2.75 M09
	Linear move **f-g** (chamfer)
	M09 Coolant off
N380	G00 X4.0 Z2.5 M05
	G00 Rapid move **g-a**.
	M05 Spindle stop
N390	T0200
	T02 Tool number 02 activate offset number 00 (00
N400	M30
	M30 End of program.
%	Rewind stop code.

SUMMARY

- CNC chucking and turning centers have evolved from standard engine lathes to high precision, computer controlled, multi-axis machines.
- Some machines can have an additional C axis, which provides fully programmable live tooling.
- The C axis provides the machine with the capability of performing milling and drilling operations.
- Four-axis machines are capable of internal and external machining operations simultaneously.

- CNC chucking and turning centers can be supplied with several optional material handling devices that improve the machine's performance and productivity.
- With a full range of G-codes, M-codes and machine cycles available, programming has become relatively simple.
- CNC chucking and turning centers once setup, run virtually unattended with a minimum of downtime making these machines exceptionally versatile.

KNOWLEDGE REVIEW

Chucking and Turning Centers

1. How were contour cuts made on the lathe prior to NC?
2. How did early NC lathes make contour cuts?
3. List five improvements found on today's CNC chucking and turning centers.

Types of Turning Centers

4. How are CNC turning centers classified?
5. What is the most common turning center used in industry?
6. What is the third axis on a three axis turning center?
7. What are two advantages of the four axis turning center?
8. Where would you use a six axis turning center?
9. What are the main parts of a CNC chucking and turning center?
10. Why are machine beds usually made of high-quality cast iron?
11. Why is the bed of chucking and turning centers usually slanted?
12. What is the purpose of the headstock?
13. What programmable spindle speeds are available on CNC turning centers?
14. Name three different types of tailstocks.
15. Why does the tailstock travel on its own bedways?
16. How is the programmable tailstock moved?
17. How is the tailstock protected against collisions with the indexing tools?
18. Name the three most commonly used types of turrets.
19. What is the repeatability of the servo system?

20. List five functions of the MCU.
21. What is the purpose of the steady rest and the follower rest?

Material Handling Devices

22. List and state the purpose of six material handling devices used with chucking and turning centers.

Machine Axes

23. What are the two primary axes on a CNC turning center?

Function Codes

24. What is a G code used for in CNC programming?
25. What do the following G codes represent in CNC programming: G00, G20, G41, G74, G96?
26. What is an M code used for in CNC programming?
27. What do the following M codes represent in a CNC program: M00, M03, M11, M17, M30?

Tool-nose Radius Compensation

28. What are the three G codes used in TNR compensation?
29. What are two effects on the workpiece of not using TNR compensation?
30. Why is the radius put on the nose of a turning tool?
31. What is the part zero location in the X axis?
32. What are the three possible locations for the Z axis part zero?

Unit 19
CAD/CAM Lathe Programming

Computer Aided Design (CAD) and Computer Aided Manufacturing (CAM) have revolutionized the mechanical design and the manufacturing process Fig. 8-19-1. The programmer no longer needs to make complicated mathematical calculations to solve problems such as center positions, intersections, tangencies, or for generating complex shapes and surfaces.

A CAD/CAM system is an integrated software tool that is capable of preparing engineering drawings, designing the workpiece, producing numerically controlled programs and simulating the cutting tool path.

With the use of the computer and the CAD/CAM software the programmer creates the geometry for manufacture of the part. The programmer should select the tools required to machine the part, the type of toolpath required for the machining operations, and the machine that will be used to produce the part. The completed CNC program is generated automatically from the information supplied by the programmer. The programmed toolpath can be viewed on the computer screen in two dimensional (2D) or three dimensional (3D) simulation to check for errors before using the program on the machine tool.

> **The one-column format used in this CAD/CAM Unit allows the placement of illustrations as close as possible to the corresponding instructions in the text. Students then can immediately and easily see if their computer screen input matches the illustrations in the text, or if a programming error has been made.**

OBJECTIVES

After completing this unit you should be able to:

1. Program a part using the CAD/CAM software provided.
2. Create a finished program for a CNC chucking center.
3. Simulate the cutting tool path.

KEY TERMS

CAD/CAM	cursor	elements
filename	geometry	icon
intersections	simulation	tangencies

Fig. 8-19-1 CAD/CAM systems have revolutionized mechanical design and manufacturing processes. *(Emco Maier Corp.)*

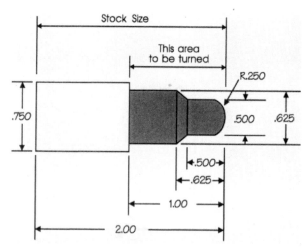

Fig. 8-19-2 A sample part that can be designed using CAD/CAM software. *(Denford, Inc.)*

4. Modify the CNC program to eliminate any errors.

DESIGNING A PART

Design engineers use CAD software to design parts, and if they are also expert in machining, they may transfer the data to CAM software to prepare the NC program. More often, the program is written by a programmer experienced in manufacturing operations

CAM software allows the engineer to design the part on a computer screen and the programmer can then convert the program by adding the appropriate G and M programming codes into a CNC program that a machine tool can understand, to manufacture the part.

There are many types of CAD/CAM software programs available and it would be impractical to attempt to explain them all. Generally most CAD/CAM programs are similar in the way they create geometry, define the cutting tools, process the data for an CNC program and simulate the cutting tool path. Once a person understands how to use a CAD/CAM software program, they are well on their way to understanding most of the standard programs used by industry.

CODES USED IN THIS UNIT	
L	= Left click (press on the mouse button)
L & H	= Left click and hold mouse button down

This textbook comes with a software program that allows a person to design a lathe program for the part shown in Fig. 8-19-2. By carefully following the steps outlined, it should be easy to program the following part.

1. Load **Lathe CAM Designer** as per the software instructions.
2. Set the **Machine Type** to **Microturn** (inches).
   — Click on the arrow next to where it says Machine Type.
   — Highlight Microturn (inches) and press L mouse button
3. Set the **Billet** (workpiece) Material to **Aluminum**.
   — To do this, click on the arrow to the left of **Billet Material**.
   — From the list, select L Aluminum.

4. The billet required is .750 diameter and 2.000 in. long. However, the section to be designed is 1.000 long.
   — Change the **Diameter** to .750 and the **Length** to 1.000 by highlighting the values and typing the correct sizes.
   — The screen should look the same as Fig. 8-19-3:

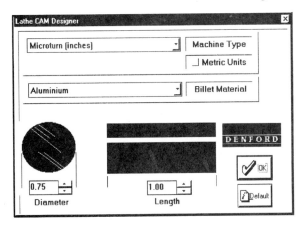

Fig. 8-19-3   Selecting the machine type and part material. *(Denford, Inc.)*

5. When all settings are correct, ☐L☐ **OK**.
   — The screen shown in Fig. 8-19-4 should now be displayed on the computer. This screen is where the geometry for the part is to be entered. The software will calculate the feedrates and depth of cut automatically.

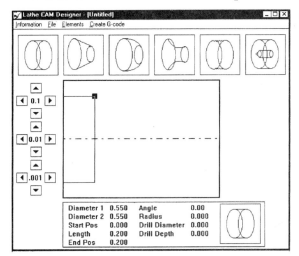

Fig. 8-19-4   The design element toolbar. *(Denford, Inc.)*

The shape elements on the top tool bar, Fig. 8-19-4, indicate the type of machining operations that can be performed by clicking on them. The elements shown in Fig. 8-19-5 will be used in entering the part dimensions.

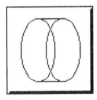

Fig. 8-19-5 Elements used to enter the shape of the part in Fig. 8-19-2 for machining operations. *(Denford, Inc.)*

The first element is a parallel section that is .375 long and .625 in. diameter, Fig. 8-19-6.

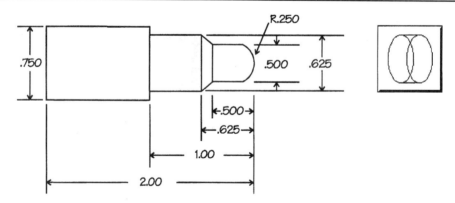

Fig. 8-19-6 The parallel element selected to enter the .625 diameter. *(Denford, Inc.)*

6. Place the cursor arrow on the small square called the **element anchor**, Fig. 8-19-7.

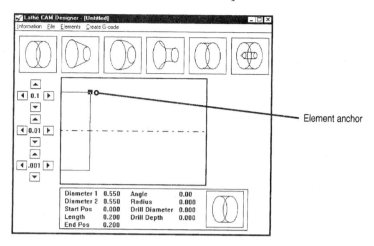

Fig. 8-19-7   The element anchor is used when programming the part. *(Denford, Inc.)*

7. L & H the left mouse button, drag the element to the top right until **Diameter 2** reads .600 and the **Length** reads .400 in the information box at the bottom of the screen, Fig. 8-19-8.
   — When the correct position is reached, release the mouse button.
8. To get the element to the desired .375 length and .625 diameter, use the **micro adjustment keys**.

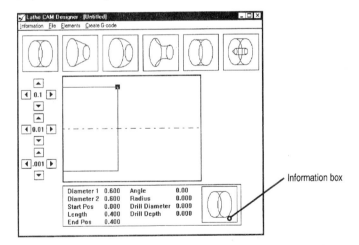

Fig. 8-19-8   The bottom information box showing the approximate diameter and length of the first section, *(Denford, Inc.)*

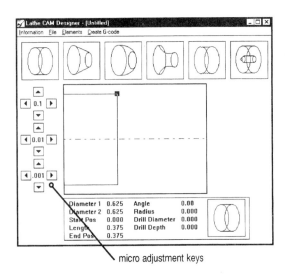

Fig. 8-19-9   The micro-adjustment keys are used to set the diameter to .625 and the length to .375. *(Denford, Inc.)*

— Place the cursor arrow on the ◀ ▶ ▼ ▲ keys for adjustments in the .100, .010, or .001 range until **Diameter 2** reads .625 and the **Length** reads .375, Fig. 8-19-9.

— The left and right adjustment keys change the **Length**

— The up and down adjustment keys change the **Diameter**.

The taper element is the second to be entered. It has a large diameter of .625, a small diameter of .500, and a length of .250.

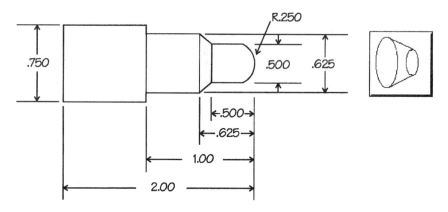

Fig. 8-19-10   The taper element selected to enter the tapered section. *(Denford, Inc.)*

9. To enter the tapered section, select the **Taper** element by clicking on its icon, Fig. 8-19-10.

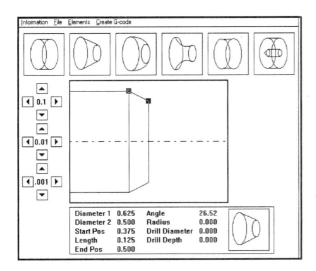

Fig. 8-19-11 Using the element anchor to enter the taper section to the proper sizes. *(Denford, Inc.)*

10. $\boxed{\textbf{L \& H}}$ the left mouse button on the **element anchor** and drag it to a **Diameter 2** of .500 and a **Length** of .125, then release the mouse button, Fig. 8-19-11.

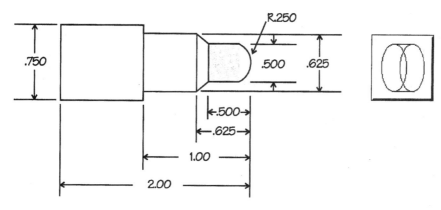

Fig. 8-19-12 Select the parallel element to enter the .500 diameter section. *(Denford, Inc.)*

> The parallel element is the third step in the process. Its dimensions are a large diameter of .500, and a length of .250.

11. Select the **Parallel** element and the screen should be as shown in Fig. 8-19-12.

12. Place the cursor arrow on the anchor and drag it to a **Diameter** of .500 and a **Length** of .200. Use the micro adjustment keys to adjust the length to .250, Fig. 8-19-13.

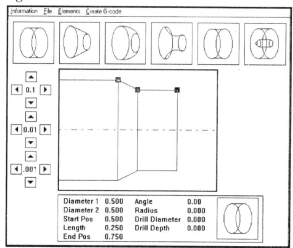

Fig. 8-19-13 Using the anchor element to adjust the .500 diameter section to the correct sizes. *(Denford, Inc.)*

To enter the last element or the convex radius on the end of the part.

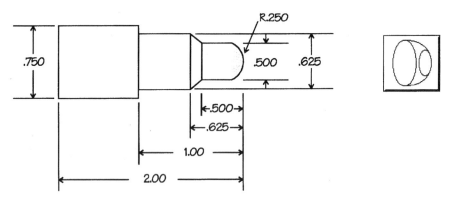

Fig. 8-19-14 Selecting the convex curve element. *(Denford, Inc.)*

13. Click the **Convex Curve element icon;** Fig. 18-19-14 shows what this icon looks like.

14. ⎡L & H⎤ the **element anchor** and drag it to a **Diameter 2** of .000 and a **Length** of .250, then release the mouse button, Fig. 8-19-15.

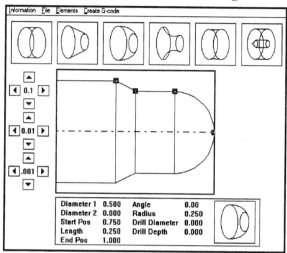

Fig. 8-19-15 Using the anchor element to create the .250 radius on the end of the part. *(Denford, Inc.)*

15. Now create a **G-code file** that the CNC lathe can read, Fig. 8-19-16.
— From the menu bar ⎡L⎤ the **Create G-code** then ⎡L⎤ the **Make file**.

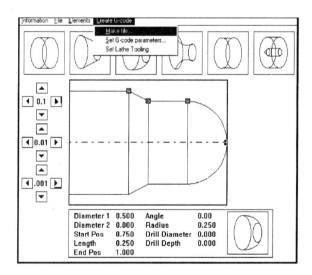

Fig. 8-19-16 Creating the G-code file. *(Denford, Inc.)*

16. Now give the program a **filename**, Fig. 8-19-17. It is a good idea to save the file with a unique name.
    — Type a filename that relates to the part that has been created.
    — Make sure the file is saved to the **c:\denford\data** directory.
Once a filename has been entered, [L] the **OK** button.

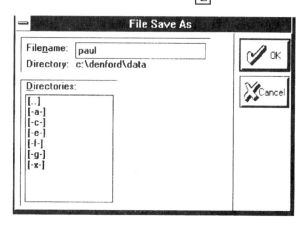

Fig. 8-19-17   Creating the program filename. *(Denford, Inc.)*

17. A box will appear showing the filename and path. [L] the **Create G code** button, Fig. 8-19-18.

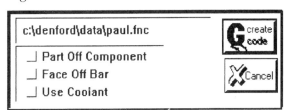

Fig. 8-19-18   Creating the G-codes. *(Denford, Inc.)*

18. The **post processor** window will now appear, Fig. 8-19-19. As the computer processes the file a line will move across the screen. When it reaches the end of the box, **L** the **OK** button.
Note: After posting is complete, the **Post Processor box** may disappear. If this happens, the file is complete.

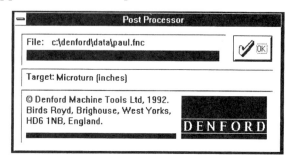

Fig. 8-19-19   The post processor screen. *(Denford, Inc.)*

19. To close **Lathe CAM Designer** select **File** from the menu bar, then select **Exit**.
20. Lathe CAM Designer will ask if the work done so far should be saved, Fig. 8-19-20. **Be sure to save the CAM file**. Prior to this the **G-code file** was saved.
    — ⎡L⎤ the **OK** button.

Fig. 8-19-20   **CAM file Save confirmation screen.** *(Denford, Inc.)*

21. A filename will now be requested, Fig. 8-19-21. Use the same filename that was used in Step 16. The software will automatically assign a **lcd extension** to the file.
    — This filename will not overwrite the file that was saved earlier because this file has a different file extension.
    — Once a filename is entered, ⎡L⎤ **OK**. Lathe CAM Designer will now exit.

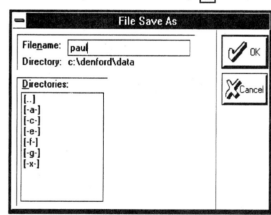

Fig. 8-19-21   **Saving the completed program.** *(Denford, Inc.)*

SUMMARY

- CAD/CAM saves time resources and production costs.
- CAD/CAM is efficient and accurate.
- With CAD/CAM software the programmer is capable of creating geometry, finished engineering drawings, graphically displaying cutting toolpaths, and editing CNC codes.
- The programmer has the ability to rotate, pan, zoom and create new views of the workpiece.
- The projected cutting tool path can be simulated in 2D or 3D and any errors can be detected and corrected before machining starts.
- The completed CNC program can be verified, printed and plotted at the computer terminal, either on the screen or as a hard copy plot.

KNOWLEDGE REVIEW

1. Define the following terms.
   a) CAD   b) CAM
2. List five advantages of CAD/CAM software.
3. Define a CAD/CAM system.
4. Name four functions that a CAD/CAM system can perform.

5. How can the simulated cutting tool path, be viewed by the programmer?
6. How does simulation aid the programmer?
7. How is the workpiece designed?

SECTION 9

CNC Manufacturing Outlook

PART I – LOOKING BACK

Any look forward has to be matched by the look back. The historical viewpoint is a necessary step to think about the present, and it is equally important to look into the future. Since its beginning in the 1950's, numerical control (NC) technology has gone through several important periods. These periods had a major effect on the metalworking industry and machining, and can be divided into three major areas:

1. Period of columns
2. Period of zeros
3. Period of decimals

In the early period, the **era of columns** covers hard-wired machine tools of the NC type (no computer numerical control - CNC in those days), and the programming format that required an input of data in a columnar form, Fig. 9-1. The position of data in a column specified which axis or function was to be performed. Letters were not allowed in the program.

In the middle period, Computerized Numerical Control became popular, and CNC machines started finding their way into medium-sized manufacturing companies. Using a letter oriented programming format (**word address programming**) with trailing zeros input was a major breakthrough in part programming. In machining, the CNC technology introduced unique program editing options, right at the operator's fingertips.

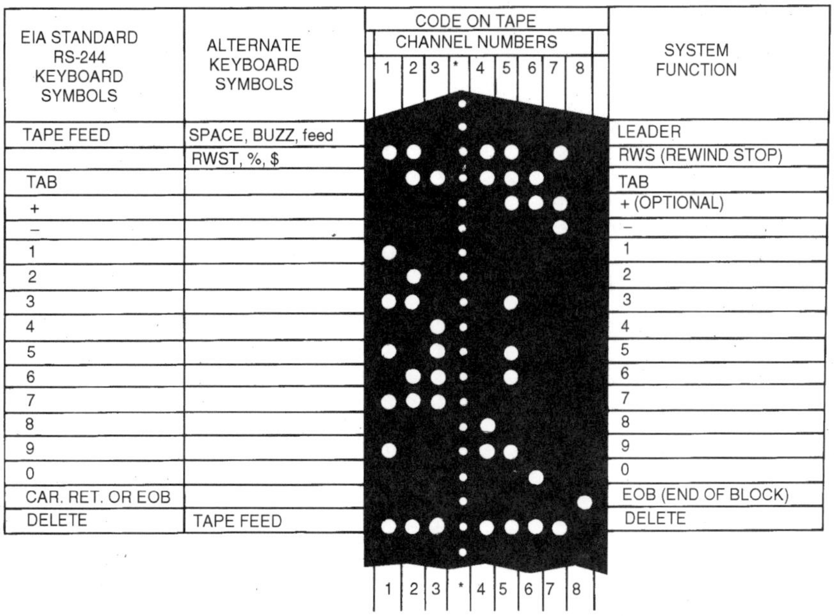

Fig. 9-1 The EIA standard NC tape consisted of eight vertical channels where holes were punched to provide coded machining information.

The latest improvement was not far behind. CNC technology has matured and preprogrammed cycles and special routines that are very easy to use have replaced many difficult programming operations of the past. Additional equipment and accessories have found their way to the manufacturing floor, helping to increase manufacturing flexibility, Fig. 9-2. In part programming, the decimal point input has become a reality and programming has found a new standard accepted by many manufacturers to this day.

A Look into the Future

The past five to ten years have also brought better software, accessories, and the present Personal Computer (PC) to the forefront. Most people in industry are quite familiar with what technology is available on the market, at least in their own fields of knowledge. With the information age, it is easy to obtain data on the latest trends, developments, and technologies. The natural question is - what will the future bring? What can be seen in this imaginary crystal ball? What can we ex-

Fig. 9-2 Modern computers, with many features not available on earlier models, have made CNC programming relatively easy. *(Emco Maier, Corp.)*

pect in manufacturing in general, and in CNC technology in particular? Is it possible to even attempt such a process?

Presenting this question to anyone knowledgeable in the particular field usually results in some sort of a rebuff. Who would want to place their reputation on the line with some possible but not always probable predictions? In a way, such a request may be unfair in the first place, for the simple reason that it tries to force an answer at any cost. That is not to say the question presented is not valid and cannot be addressed in some other way.

There are certain trends that may or may not turn out the way we think about them today. That may be disappointing to some, but every trend that exists today is a potential seed that can grow, given the proper nurturing. Every trend may be a start of something new, and a new trend may start the human thinking process well before the actual physical development phase can take place.

Advanced Manufacturing

Advanced manufacturing is a very large and complex concept, and CNC is only a very small part of it. The main idea of advanced manufacturing is to build on existing technologies and to adapt new and advancing technologies that enable industries to increase their productivity and their overall capabilities. Advanced manufacturing policy also provides fertile ground for expansion of new product line or services. To be successful, the initial policy has to become the main philosophy of management. Implementation of new technologies has to be company or even corporation wide, not just consisting of small pockets of improvement. Therefore, it is impossible to attempt a look at the CNC future, without looking at the larger picture. This is – in effect – the purpose of this unit. The idea is to provide a look at technologies that influence CNC to some extent - technologies that are related to CNC and may have an effect on it, even if they are not used directly.

Many professionally conducted independent studies have identified several leading technologies of the present that will have a great influence on the future development in manufacturing, Fig. 9-3

Each item will be covered briefly although a book could be written on each topic to provide a comprehensive overview that lack of space within this section will not allow.

PART II – ADVANCED TECHNOLOGIES

We have only barely scratched the surface of what computers can do for us, even if used properly. The ambitious dreams of the nineteen fifties may not all have materialized; yet they certainly contributed to the effort to use computers in a way that benefits the consumer or the client and the manufacturer. All advanced technologies used in manufacturing depend on the use of computers. This is a growth area, as better and more improved hardware and software is becoming available, literally on a daily basis.

Computerized Numerical Control (CNC)
The word **Computerized** in the **Computerized**

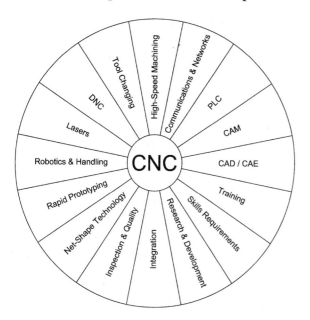

Fig. 9-3 Some of the present technologies that will have a great influence on manufacturing in the future. *(Peter Smid)*

Numerical Control has been around since the middle 1970's, and this technology will continue to play an important factor in manufacturing. More and more machine controls are becoming PC based, currently using Windows as the main operating system, often combined with PLC (programmable logic control) software controlling various phases of automation. This combination opens up many possibilities and opportunities.

Several new technologies will certainly find their way into mainstream manufacturing (CNC included); namely those referred to as **simulation**, **virtual reality**, and so on. The Computer Aided Design and Engineering concept has been around for quite a while, but it had always lacked the feature of showing the designed part working with other parts in assembly or manufacturing flow. The simulation and virtual reality technology closes this gap.

On a simple level, there are many program editors available in CNC work that simulate the tool path of a part program, often only using general vector type graphics for back plotting, Fig. 9-4. That may be an improvement from earlier years, and it is only the beginning of something much more sophisticated. True simulation software will allow not only the tool path to be seen on the computer screen, but also a tool path using a solid part model, including the fixtures, the tooling, and even the machine tool and its motions. As all motions can be made available, this virtual machining center or lathe will offer much more realistic representation of the machining process. The benefits of catching errors before they do any damage will result in more productive machining and, indeed, in manufacturing overall. Of course, the capabilities are much higher than that.

Computer Aided Design and Engineering
The field of engineering design has experienced a very fast growth rate and this growth is expected to continue well into the future. In their most general form, CAD and CAE are used for designing parts for manufacturing, using computers and suitable software. These technologies

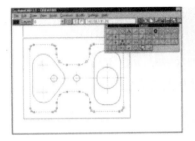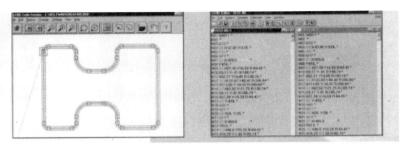

Fig. 9-4   Modern CNC program editors allow the simulation of a cutter tool path and the editing of the program. *(Denford, Inc.)*

are also used to combine with the actual process of manufacturing, for example by using a design as the only source for a CNC machining program. Designs made electronically have replaced the traditional storage methods, such as paper drawings. They are suitable for exchange between companies, for example from the designer to the manufacturer, and from the manufacturer to different departments.

The latest trends in the CAD/CAE technologies are also new related technologies, namely part visualization through a development of a computer model and a process simulation. This area of virtual manufacturing (and machining) is certain to grow rapidly. It brings to mind the example of early 1990's of a Boeing 767 aircraft developed entirely from computer models, abandoning traditional methods used before.

Computer Aided Manufacturing

Computer Aided Manufacturing (CAM) has been a direct extension of the Computer Aided Design (CAD), resulting in the acronym CAD/CAM. Although many CNC machinists use the word CAM as synonymous with the development of CNC programs using a computer and software, the term covers a much larger area. In fact, all items mentioned in this chapter belong to the category of Computer Aided Manufacturing.

Programmable Logic Controllers

Machining systems will also grow, particularly in mass production industries such as the automotive sector. CNC (Computerized Numerical Control) technology combined with the use of PLC (Programmable Logic Control) can make CNC machining part of a larger process, such as production lines, often referred to as transfer lines. PLC is a specially designed system that is used to control switching of various devices, and is used in all phases of automation.

Computer Communication Networks and Information Exchange

As the saying goes, information is knowledge. Having accurate information available on demand, instantly, will be the great challenge of the future. Internet and Intranet will be a major source of data, as more information will be available in the form of tables, charts, databases, catalogs, company profiles, services offered, direct selling, auctions, etc.

Just as computers on every desk are getting to be reality now, the computer communications networks we know as the Internet or the Intranet is growing at an undreamed of speed, Fig. 9-5. The current focus of just **browsing** the Net will turn into a strong research and marketing tool, where having a presence on the Net in the future will become more important than having a business card or a cell phone today. Internet will also be more accessible from mobile phones and available in more public places, such as hotels and convention centers. Internet offers instant access to the world markets and comes complete with words, pictures, animation, sounds, links and other features. Communications network offers the possibilities of maintaining business contacts with customers, clients, suppliers, vendors, subcontractors, head offices, etc.

More companies will be using the Internet (or

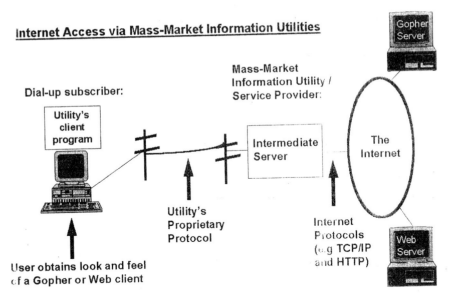

Internet Access via Mass-Market Information Utilities

Gopher Server

Dial-up subscriber:

Utility's client program

Mass-Market Information Utility / Service Provider:

Intermediate Server

The Internet

Utility's Proprietary Protocol

Internet Protocols (e.g TCP/IP and HTTP)

Web Server

User obtains look and feel of a Gopher or Web client

Fig. 9-5 The Internet provides easy access to world markets, research, and marketing. *(McGraw-Hill, Inc.)*

an Intranet within the company) as a tool for training and education, customer support, advertising and marketing, accounting and financial transactions, research, exchange of ideas, exchange of documents, and general correspondence (E-mail) with suppliers and customers. This type of local area network can be useful in exchanging information between various departments, for example, between engineering and production control, or between production control and the manufacturing shop.

High-speed Machining

In manufacturing and in the CNC metal cutting field specifically, the trends appears to be higher machining speeds. Machine tools have already passed the 1000 in/min rapid traverse rate, and there are machines approaching the 2000 in/min level. The main benefit of rapid rate is to reduce unproductive time during machining. Cutting motion speed has not lagged behind either, and high-speed machining centers are not uncommon anymore. Cutting rates are and will continue to be higher than rapid traverse rates were just a few short years ago. Today, spindle speeds on high-speed machining centers start at 10,000 to 15,000 rev/min,

and much higher speeds are used in practical applications. High-speed machining benefits any industry where 3D (three-dimensional) machining is required, such as the mold and aerospace industries. Its main benefit is the elimination of after-machining operations, such as bench work. Even where bench work is necessary, it is kept to the minimum.

Tool Changing

Another development that can be expected to improve manufacturing productivity is improved automatic tool changers on CNC lathes, Fig. 9-6. This emerging design will incorporate tool storage away from the machining area, similar to that of CNC machining centers. This trend is already well underway but will grow rapidly in the future, as more ideas find their places in the market. CNC lathes will continue to incorporate more milling features by improving on the technology of live tooling available today.

Direct Numerical Control - DNC

DNC has been around since the early days of CNC. Its primary purpose has been to upload and download CNC programs to and from the memory area of the CNC machine control. One

Fig. 9-6 Automatic tool changers on CNC machine tools increase manufacturing productivity by having tools ready in a storage magazine. *(Cincinnati Machine, A UNOVA Co.)*

or more machines can be reached from a central computer, giving DNC its second meaning, **Distributed Numerical Control**. This trend will most certainly continue, but will be complemented by the rise in the ability of many software programs to offer so called **drip-feed**, frequently used in mold-making industries, where programs are too large to fit the maximum capacity of the machine control memory. Drip-feeding uses an inexpensive microcomputer, and bypasses the memory of the CNC unit altogether.

In some applications, the DNC system can be combined with a part-management system, where the CNC operator selects the latest version of the drawing and part program from a central computer console, conveniently located in the machine or manufacturing shop.

Lasers in Manufacturing

Laser technology has been used for some time in at least two areas of manufacturing – machining, and assembly calibration. Lasers will play an even more important role in manufacturing while enhancing the automated manufacturing process, Fig. 9-7. Although they can be used for metal cutting and precision welding, lasers can also be used for material marking, where other methods would be impractical or even impossible. Lasers will also be used more prominently in marking and identifying tools and finished parts. This identification can be made readable by humans or by machines in various stages of material or part handling.

Robotics and Material Handling

The robot as a production tool – the word was coined by the Czech playwright Karel Capek in the 1930's play **R.U.R.** – is widely used. Robotic technology today is a sophisticated area within the manufacturing process. Often associated

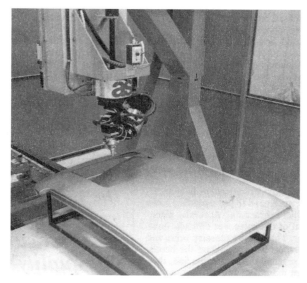

Fig. 9-7 Lasers are presently used in manufacturing for measurement, metal cutting, welding, and material identification. *(Manufacturing Engineering Magazine)*

with automatic welding and assembly (as well as bomb disposal police units), robots can be used for much more complex tasks, Fig. 9-8. The simpler robotic tasks use preprogrammed motions through the method of **teaching**. Once taught, the robot can repeat the **learned** motions accurately any number of times. Advanced use of robots will include sensors, vision systems, and

Fig. 9-8 The robots of tomorrow will include improved sensors, vision systems, and artificial intelligence to make them a valuable production tool.
(The Association for Manufacturing Technology)

similar devices that will be able to change the robot activity, based on a certain conditions, as they develop.

Robots are not the only technology in material and part handling. Gantry systems, as an integral part of mass production, are suitable as excellent replacement for cranes and hoists. Automatically guided vehicles can eliminate manual transportation of parts between machines, Fig. 9-9. Agile manufacturing combines many technologies into what is often called **lean manufacturing**. Such automation frequently requires sophisticated methods of part identification to be in place. Each part has to be accounted for during all machining processes and for storage and shipping purposes. Uses of laser technology, described earlier, aid in this effort.

Rapid Prototyping

Rapid prototyping technology is sometimes known as three-dimensional plotting (3D plotting). Its main advantage is the capability to develop a layered plastic prototype part directly from a design model of a CAD/CAM system, Fig. 9-10. Such a part can be a full size or a precisely scaled model. The benefits of rapid prototyping are mainly applied in the related technology of net shapes and near-net shapes. The combination of rapid prototyping and net-shape processes in castings eliminates problems associated with sand casting porosity and as a result reduces manufacturing costs.

Net-shape and Near Net-shape Technologies

This is one of the relatively newest advanced technologies that will have a great effect on manufacturing. It is based on the investment-casting technology, where special materials cannot be formed by other methods. Mainly used in the die casting and metal forming industries, the technology offers manufacturing of completed parts with minimal machining. Its main purpose is to eliminate, or at least to reduce, machining costs. Often, no machining is required at all, as this technology supports very high tolerances that can be applied to the casting. A related operation,

Fig. 9-9 Modern automatic guided vehicles will reduce the manual transportation of parts in a manufacturing operation. *(Giddings & Lewis, Inc.)*

called fineblanking, is a stamping press process that can produce parts up to .750 in. thick to within .001 in. accuracy with 8 to 20 microinch surface finishes, Fig. 9-11.

Inspection and Quality

Any continuous improvement in the production process depends on the control of part quality. The old methods of inspecting a part after so many parts have been made, does not work with advanced technologies anymore. The idea of eliminating scrap parts from production has been replaced by the idea of preventing scrap in the first place. The new technologies of inspection, all computer controlled, must identify potential problems during manufacturing, before any damage is done. Computer-controlled sensors and vision equipment can be used to measure the part and distinguish parts that are within tolerance specifications and out of tolerance. Broken tools, improperly set cutting tools, part dislocated in the fixture, bottlenecks on the line, and hundreds of other obstacles can be detected. In related applications, various sensing devices can be used to monitor production process and identify problem areas. Use of optical-sensing devices, video cameras and similar devices will

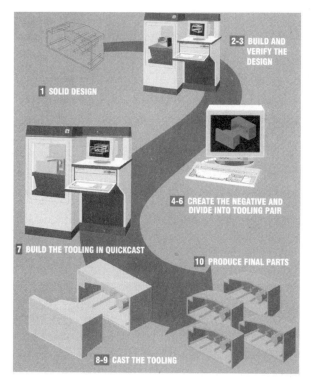

Fig. 9-10 Rapid prototyping can develop a layered plastic prototype part directly from the CAD/CAM system design model. *(3D Systems)*

Fig. 9-11 The fineblanking process can produce stamped metal parts up to .750 in. thick to within .001 in. accuracy. *(Feintool Cincinnati, Inc.)*

definitely play a major role in continuous improvement.

Integration

Any technology has been developed for a particular purpose, often for a single purpose only. For example, CNC machining has developed as means to automate the machining process and increase part quality and manufacturing cycle time. However important individual technology may be in its own right, it cannot match its importance as part of a larger manufacturing concept.

In terms of computers and manufacturing, the main purpose of integration is to control the process of manufacturing. MRP (**Manufacturing Resource Planning**) is the most common element, monitoring all key aspects of manufacturing, such as production scheduling, loading of machines, material flow, part flow, etc. Many common activities and individual functions of the manufacturing process can be controlled by computers and offer a greater degree of automation. CIM (**Computer Integrated Manufacturing**) is the name of a technology that has been used for such purpose. Apart from the process control, some of its elements include machine monitoring, control status, data collection, various aspects of remote control operations, material handling, robot operation, etc.

Research and Development

An often-neglected aspect of many companies is in the area of research and development (R&D), Fig. 9-12. A common reason is that these companies do not manufacture a unique product. Many companies forget or even ignore the reality that the purpose of R&D is not only to develop new products and make current products better, but also to improve the overall process of doing business. This other reason is common to all companies and should not be taken lightly.

Fig. 9-12 Research and development must remain an important factor in the survival of a company. *(The Association for Manufacturing Technology)*

PART III SKILLS REQUIREMENTS OF THE FUTURE

Regardless of how many computers one uses, regardless of how computerized a company is, there will always be requirement for people with specific skills. Today, we may call these skills **hi-tech**, tomorrow they will be ordinary skills. Demand for these advanced technology skills will be increasing at the same time as the technology changes. It is only normal to see some skills disappear just as it is normal to expect increased need for new skills. Any change in technology comes with changing requirements for skilled people.

The typical areas of skills that will grow belong to several categories, namely computer skills of all kinds, various technical skills in manufacturing and service sectors, including CNC programming and CNC machining, PLC programming, and quality control. Common to all skill requirements will be need for good communicators, as more trade and commerce will take on global form.

It is generally expected that there will be shortage of some skilled personnel in the years to come, which will slow down the introduction of advanced technologies. In the area of skilled trades, the demand for CNC machinists and CNC operators is expected to grow, particularly for those professionals who can make their own CNC programs and setups. Related to these CNC skills, more traditional skills will also be in high demand, particularly in the area of tool and die work and mold making. High technology industries, such as electronics and aerospace, will require workers with multi skills and problem solving capabilities.

Not only will front line workers be affected, but demand for managerial skills in the same industries will also be high. As manufacturing accepts new and advanced technologies, production management skills will undoubtedly also be redefined. That change will have a direct effect on the management of design engineering and even human resources.

In the area of computers, demand for software development skills (computer programming) will be high, along with demands for skills in administration and management of the information network system and computer hardware and peripherals.

Training

Properly conducted and focused training is also part of the continuous improvement process. Companies should look toward two resources, external and internal. The traditional players in the external group of resources, namely schools, colleges, universities, skills training institutions and, to some extent, vendor training, will play a large role. Those that do not practice the philosophy of continuous improvement themselves will be not be qualified to offer anything substantial and eventually will be left out of the process.

Internal training resources are equally important, but should be focused differently. Their main focus should be training that is not or cannot be offered externally. This situation often involves various equipment vendors, vendors who have to improve their poor image as quality trainers. Making or knowing a piece of equipment does not guarantee the ability to teach how to use such equipment. Vendors have to aim at higher standards in the training area.

Summary

No technology can exist by itself. It must become part of a plan, part of a program, and part of a philosophy. Simply, advanced technology has to become part of the bigger picture, not the bigger picture itself. The main issues that complement the introduction of new and advanced technologies are the practice of continuous improvement within the company. Continuous improvement must not be a mere slogan or an empty phrase in the company mission statement. It must be a true way the company does business. Improvement of product or service quality is the first criterion, combined with just-in-time deliveries and addressing workers' issues in a forward way.

PART IV – EXPECTATIONS AND ISSUES OF CONCERN

It is only reasonable to expect certain returns from the growth of advanced technology. A typical company that adopts new technology and manages it successfully will experience, over a period of time, increased profit margins, better product or service quality, and improvement in productivity. Such a company will be in the position to offer better working environment for its employees and, in turn, expect its own people to do more for the company and themselves.

That is not to say that obstacles will not occur. Manufacturing industry, as well as other industries, always faces certain difficulties that stop or slow down faster growth. As usual, there are always several reasons, the most common obstacle being the high cost of capital and equipment. Another important problem many companies try to deal with is the shortage of skilled people at various levels. It will be a simple matter of equation, that if the disadvantages of adopting new technologies are stronger than their potential benefits, it is impossible to make any advances. If these difficulties cannot be resolved in a timely way, such a company will face much slower growth and may even experience a reverse trend.

Not all solutions to these concerns are within the capability of any particular company. Any change must originate within such a company, even it means drastically changing established practices as how to run a business. Advanced technology of any kind is not a solution in itself; it becomes only means to reach a previously established goal.

Glossary

A axis - The axis of rotary motion of a machine tool member or slide about the X axis.

absolute dimension - A dimension expressed with respect to the initial zero point of a coordinate axis.

absolute programming – A mode of CNC programming in which all axis movements are made in relation to a fixed datum point.

- The preparatory function G90 must be included in the CNC part program for absolute programming, Fig. G1.

absolute readout - A display of the true slide position as derived from the position commands within the control system.

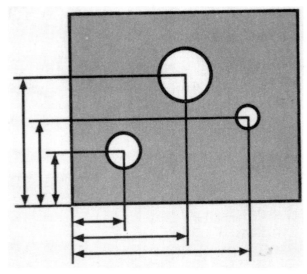

Fig. G1 Absolute programming.
(Cincinnati Milacron, Inc.)

absolute system – A CNC system in which all positional dimensions, both input and feedback, are measured from a fixed point of origin.

acceleration and deceleration (ACCENDEC) - Acceleration and deceleration in feed rate, providing smooth starts and stops when a machine is operating under CNC and changes from one feed rate value to another.

access time - The time interval between the instant at which information is: 1. Called for from storage and the instant at which delivery is completed, i.e., the read time. 2. Ready for storage and the instant at which storage is completed, i.e., the write time.

accuracy -
1. Measured by the difference between the actual position of the machine slide and the position demanded.
2. Conformity of an indicated value to a true value, i.e., an actual or an accepted standard value. The accuracy of a control system is expressed as the deviation or difference between the ultimately controlled variable and its ideal value, usually in the steady state or at sampled instants.

active storage – The part of the control logic that holds the information while it is being transformed into motion.

ADAPT - An Air Force adaptation of APT, which has limited vocabulary and can be employed on some small- to medium-sized U.S. computers for CNC programming.

adaptive control – A means of automatically adjusting feeds and/or speeds of a cutting tool from sensor feedback (continuous monitoring) to maintain the best cutting conditions, Fig. G2.

- Sensors may measure cutting forces, torque, cutting temperature, vibration amplitude, horsepower or spindle deflection, work material hardness, and width or depth of cut.
- Adaptive control has the capability to respond to and adjust for these variations during machining.
- It also provides long tool life and/or lower machining cost.
- Current adaptive control machining systems are generally systems where certain limits are set on each process variable.

ALGOL - (see algorithmic language)

algorithm - A rule or procedure for solving a mathematical problem that frequently involves repetition of an operation.

algorithmic language (ALGOL) - Language used to develop computer programs by algorithm.

alphanumeric - A system in which the characters used are letters A through Z and numerals 0 to 9.

American Standard Code for Information Interchange (ASCII) - A data transmission code that has been established as a standard by the American Standards Association. It is a code in which 7 bits are used to represent each character. (Also known as USACII.)

analog - In CNC, the term applies to a system that utilizes electrical voltage magnitudes or ratios to represent physical axis positions.

analog-to-digital (A/D) converter - A device that changes physical motion or electrical voltage into digital factors, Fig. G3.

Fig. G2 Adaptive control. *(Cincinnati Milacron, Inc.)*

APT - (see automatically programmed tools)

arc clockwise - An arc generated by the coordinated motion of two axes in which curvature of the path of the tool with respect to the workpiece is clockwise when the plane of motion is viewed from the positive direction of the perpendicular axis.

arc counterclockwise - Same as arc clockwise above, except substitute "counterclockwise" for "clockwise."

ASCII - (see American Standard Code for Information Interchange)

automatically programmed tools (APT) – A universal computer-assisted program system for multiaxis contouring programming APT III. Provides for five axes of machine tool motion.

automatic system for positioning of tools (AUTOSPOT) – A computer-assigned program

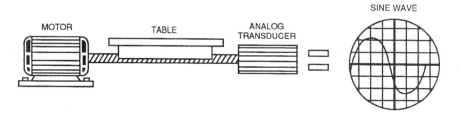

Fig. G3 Analog to Digital convertor. (*Coleman Engineering Co.***)**

for CNC positioning and straight cut systems, developed in the United States by the IBM Space Guidance Center. It is maintained and taught by IBM.

automatic tool changer (ATC) – A mechanical arm-like device on the CNC machine tool for automatically changing cutting tools under the control of a part program, Fig. G4.

automation - The technique of making a process or system automatic.
- Automatically controlled operation of an apparatus, process, or system, especially by electronic devices.
- In present-day terminology, the term is usually used in relation to a system whereby the electronic device controlling an apparatus, or process also is interfaced to and communicates with a computer.

AUTOSPOT - (see automatic system for positioning of tools)

auxiliary function - A function of a machine other than the control of the coordinates of a workpiece or cutter - usually ON/OFF type operations. These functions are typically the MST (Miscellaneous, Spindle, Tool) functions.

axis - 1. A principal direction along which a movement of the tool or workpiece occurs. 2. One of the lines of reference of a coordinate system, Fig. G5.
- The common axes used on machine tools are:
  Primary axes X, Y, Z (milling machine)
  X, Z, (lathe)

axis inhibit - Prevents movement of the selected slides with the power on.

axis interchange - The capability of inputting the information concerning one axis into the storage of another axis.

axis inversion - The reversal of normal plus-and-minus values along an axis, which makes possible the machining of a left-handed part from right-handed programming or vice versa.
- Also known as mirror image.

B axis - The axis of rotary motion of a machine tool member or slide about the Y axis.

backlash - A relative movement between interacting mechanical parts, resulting from looseness.

batch processing - A manufacturing operation in which a specified quantity of material is subject to a series of treatment steps.
- Also, a mode of computer operations in which each program is completed before the next is started.

baud - A unit of signaling speed equal to the number of discrete conditions or signal events per second: 1 bit/s in a train of binary signals and 3 bits/s in an octal train of signals.

Fig. G4 Automatic tool changer. (*Giddings & Lewis, Inc.*)

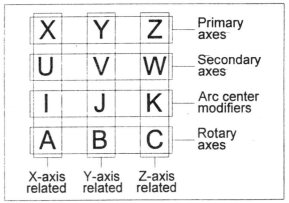

Fig. G5 Axis. (*Peter Smid*)

BCD - (see binary-coded decimal)

binary - A numbering system based on 2. Only the digits 0 and 1 are used when written.

binary-coded decimal (BCD) - A number code in which individual decimal digits are each represented by a group of binary digits.

- In the 8-4-2-1 BCD notation, each decimal digit is represented by a four-place binary number, weighted in sequence as 8, 4, 2, and 1.

binary digit (bit) - A character used to represent one of the two digits in the binary number system, and the basic unit of information or data storage in a two-state device.

bit - Either of the digits 0 or 1 when used in the binary numeration system. The smallest unit of computer data.

block - A set of words, characters, digits, or other elements handled as a unit.

- On a punched tape, a block consists of one or more characters or rows across the tape that collectively provides enough information for an operation.
- A "word" or group of words considered as a unit separated from other such units by an "end of block" (EOB) character, Fig. G6.

block skip/block delete - Permits selected blocks of the program to be ignored by the control system at the discretion of the operator with permission of the programmer.

boolean algebra - An algebra named for George Boole. This algebra is similar in form to ordinary algebra, but with classes, propositions, yes/no criteria, etc., for variables rather than numeric quantities.

- It includes the operator's AND, OR, NOT, EXCEPT, and IF THEN.

buffer storage -

1. A place for storing information in a control for anticipated transference to active storage. It enables a control system to act immediately on stored information without waiting for a tape reader.
2. A register used for intermediate storage of information in the transfer sequence between the computer's accumulators and a peripheral device.

byte - A sequence of eight bits, usually less than a word, treated as a unit and representing a character.

C axis - The axis of rotary motion of a machine tool member or slide about the Z axis, Fig. G7.

CAD - (see computer-aided design)

CAM - (see computer-assisted manufacturing)

cancel - A command that will discontinue any canned cycle or sequence commands.

canned cycle – (See fixed cycle)

Cartesian coordinates – A set of three numbers that define the location of a point within a rectangular coordinate system, which consists of three axes (X, Y, and Z) perpendicular to each other, Fig. G8. The numbers represent distances from the origin, the intersection of the three axes.

cathode-ray tube (CRT) - A display device in which controlled electron beams are used to present alphanumeric or graphical data on a luminescent screen.

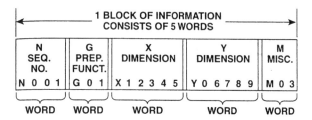

Fig. G6 **Block of information.** *(Kelmar Associates)*

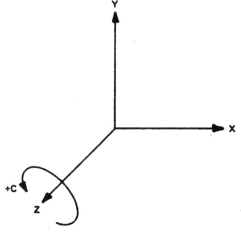

Fig. G7 **C axis.** *(Kelmar Associates)*

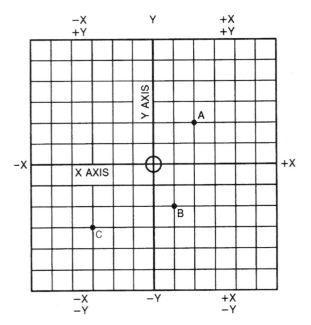

Fig. G8 **Cartesian coordinates.** *(Allen Bradley Co.)*

central processing unit (CPU) - The portion of a computer system consisting of the arithmetic and control units and the working memory.

channel - A communication path.

character - One of a set of symbols.
- The general term to include all symbols such as alphabetic letters, numerals, punctuation marks, mathematics operators, etc.
- Also, the coded representation of such symbols.

Chip - A single piece of silicon that has been cut from a slice by scribing and breaking. It can contain one or more circuits but is packaged as a unit.

circular interpolation – An interpolation scheme for programming a circular arc by specifying the coordinates of its endpoints, the coordinates of its center, and its radius, as well as the direction of the cutter along the arc, Fig. G9.
- A tool path consists of a series of straight-line segments, with the segments being calculated by the interpolation module rather than the programmer.
- The cutter is directed to move along each line segment one by one in order to generate the smooth circular path.

clear - To erase the contents of a storage device by replacing the contents with blanks or zeros.

clock - A device that generates periodic synchronization signals and regulates the operations of a processor.

closed loop - A signal path in which outputs are fed back for comparison with desired values to regulate system behavior.

compact disks – Disks holding a large amount of digital data that has been carved into them with a laser, Fig. G10.
- **CD** (compact disk) holds up to 650 MB of computer data.
- **DVD** (digital versatile disk), a double-sided optical disk that holds 750% more than CDs and can handle data, audio, and video formats.

computer - A device capable of accepting information in the form of signals or symbols,

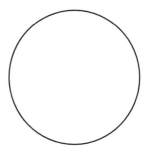
A - TRUE "CIRCLE"

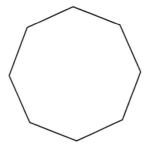
B - EIGHT SEGMENT "CIRCLE"

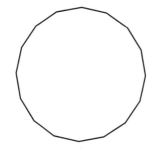
C - SIXTEEN SEGMENT "CIRCLE"

Fig. G9 **Circular interpolation.** *(Allen Bradley Co.)*

Fig. G10   Compact disk.

performing prescribed operations on the information, and providing results as outputs.

computer-aided design (CAD) - A process that uses a computer to assist in the creation or modification of a design, Fig. G11.

computer-assisted manufacturing (CAM) - A process using computer technology to manage and control the operations of a manufacturing facility.

Fig. G11   Computer-aided design. (*AMT—The Association for Manufacturing Technology*)

computer numerical control (CNC) - A method of controlling a machine/tool path change where a dedicated, stored program computer is used to perform some or all of the basic CNC functions.

constant surface speed (CSS) - The ability of a turning center MCU to maintain a constant surface speed at the cutting tool point regardless of the changes in work diameter.

continuous path operation - An operation in which rate and direction of relative movement of machine members is under continuous numerical control. There is no pause for data reading.

contouring - An operation in which simultaneous control of one or more axis is accomplished.

contouring control system - A CNC system for controlling a machine (milling, drafting, etc.) in a path resulting from the coordinated simultaneous motion of one or more axes.

coordinate system in CNC – A standard axis system that describes the positions of the cutting tool with respect to the workpiece, Fig. G12. There are three primary *linear* axes – X, Y, and Z – and three *rotational* axes – A, B, and C.

- The X and Y axes are defined in the plane of the table; the Z axis is perpendicular to this plane, and movement in the Z direction is controlled by the motion of the spindle.
- The A, B, and C axes are used to specify angles and rotation about the X, Y, and Z axes.

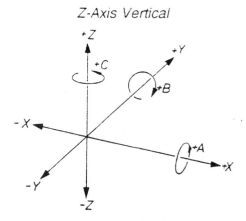

Fig. G12   Coordinate system in CNC. (*Courtesy Allen Bradley*)

- To identify positive from negative angular motions, the right-hand rule can be used, Fig. G13.

CPU -(see central processing unit)

CRT -(see cathode-ray tube)

CSS -(see constant surface speed)

cursor - A visual movable pointer used on a CRT by an operator to indicate where corrections or additions are to be made.

cutter diameter/radius compensation - A system in which the programmed path may be altered to allow for the difference between actual and programmed cutter radius.

cutter offset - 1. The distance from the part surface to the axial center of a cutter. 2. A CNC feature that allows an operator to use an oversized or undersized cutter.

cutter path/tool path – The path taken by the center of a rotating cutter or tool or the tip of a lathe tool, Fig. G14.

Cycle -
1. A sequence of operations that is repeated regularly.
2. The time it takes for one such sequence to occur.

cycle time - The period required for a complete action. In particular, the interval required for a read and a write operation in working memory, usually taken as a measure of computer speed.

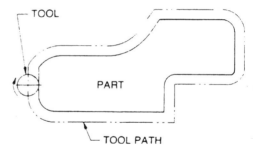

Fig. G14 Cutter/tool path. *(Allen Bradley Co.)*

datum – A reference point from which movement or measurements are made, Fig. G15.

debug - To detect, locate, and remove mistakes from computer software or hardware.

decimal code - A code in which each allowable position has 1 of 10 possible states. (The conventional decimal-number system is a decimal code.)

decoder - A circuit arrangement that receives and converts digital information from one form to another.

digital - Representation of data in discrete or numerical form.

digital computer - A computer that operates on symbols representing data by performing arithmetic and logic operations.

digital-to-analog (D-A) conversion - Production of an analog signal whose instantaneous magnitude is proportional to the value of a digital input.

digitizer - A unit that tracks the relative position

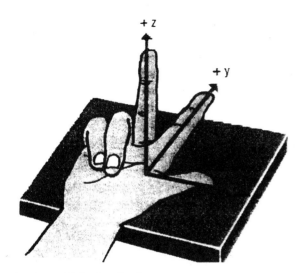

Fig. G13 Right-hand rule.

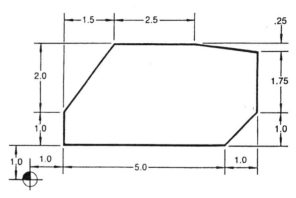

Fig. G15 Datum. *(Kelmar Associates)*

of a cursor, for the purpose of recording relative locations of a part print or actual part.

direct numerical control (DNC) – The direct control of a number of separate CNC machine tools by a central host computer. The part programs are down-loaded from a host computer directly into the memory of a CNC machine tool as required, Fig. G16.

- The system consists of four components: central computer, bulk memory, which stores the CNC part programs, telecommunication lines, and machine tools.
- The advantages of DNC are timesharing, greater computational capability for functions such as circular interpolation, remote computer location, elimination of tapes and tape reader for improved reliability, and elimination of hard-wired controller unit on some systems.

disk – A thin, round, flat magnetic data medium that is used to read or write data.

disk memory - A nonprogrammable, bulk-storage, random-access memory consisting of a magnetizable coating on one or both sides of a rotating thin circular plate.

display - Lights, annunciators, numerical indicators, or other operator output devices at consoles or remote stations.

DNC - (see direct numerical control)

documentation – The group of techniques necessarily used to organize, present, and communicate recorded specialized knowledge.

downtime - The interval during which a device is inoperative.

dump - To copy the present contents of a memory onto a printout or auxiliary storage.

dwell - A programmed pause (delay) in the processing of a CNC program.

edit - To modify a program, or alter stored data prior to output.

Electronics Industries Association (EIA) standard code - Standard codes for positioning, straight-cut, and contouring control systems.

encoder - An electromechanical transducer that produces a serial or parallel digital indication of mechanical angle or displacement.

end of program – A CNC miscellaneous function (M30) indicating the end of a part program. It stops all processing commands.

EOB character - (see end-of-block character)

error signal - Difference between the output and input signals in a servo system.

executive - Software that controls the execution of programs in the computer, based on established priorities and realtime or demand requirements.

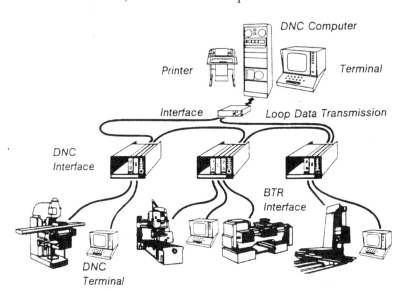

Fig. G16 Direct numerical control. *(Modern Machine Shop)*

feedback - The signal or data fed back to a commanding unit from a controlled machine or process to denote its response to the command signal.
- The signal represents the difference between actual response and desired response that is used by the commanding unit to improve performance of the controlled machine or process, Fig. G17.

feedback device - An element of a control system that converts linear or rotary motion to an electrical signal for comparison to the input signal, e.g., resolver, encoder.

feed engage point - The point where the motion of the Z axis changes from rapid traverse to a programmed feed (usually referred to as the R dimension).

feed function - The relative motion between the tool or instrument and the work due to motion of the programmed axis or axes.

feed override - A variable manual control function directing the control system to reduce or increase the programmed feedrate.

fixed block format - A format in which the number and sequence of words and characters appearing in successive blocks is constant. (obsolete on modern machines)

fixed cycle - A preset sequence of events initiated by a single CNC command: for example, G84 for CNC tap cycle. Also known as canned cycle.

fixed sequence format - A means of identifying a word by its location in a block of information. Words must be presented in a specific sequence, and all possible words preceding the last desired word must be present in the block (obsolete on modern machines)

fixed zero – The origin of a CNC coordinate system that is always located at the same position on the machine tool from which all machine movements are referenced, Fig. G18.
- Usually, it is the upper right corner of the machine table, and all locations must be defined by positive X and Y coordinates from the fixed origin.
- With fixed-zero systems, the part programmer and machine operator must reference the job to the machine's permanent zero point.

floating zero - A characteristic of an MCU that allows the zero reference point to be set at any point of the machine table travel.

format classification - A means, usually in an abbreviated notation, by which the motions,

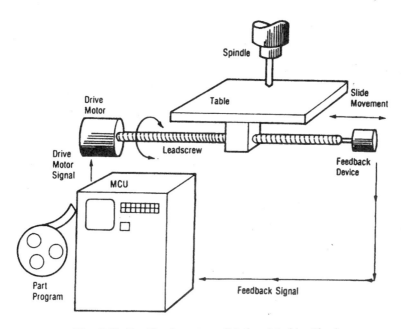

Fig. G17   Feedback system. (*Modern Machine Shop*)

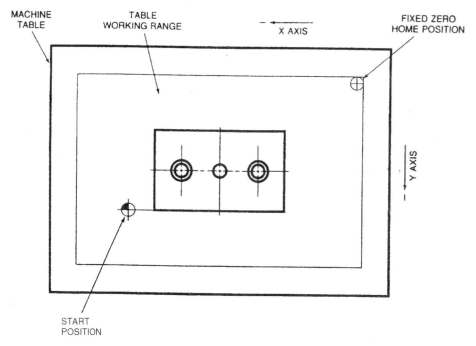

Fig. G18 Fixed zero. *(Kelmar Associates)*

dimensional data, type of control system, number of digits, auxiliary functions, etc., for a particular system can be denoted.

formula translator (FORTRAN) - An algebraic procedure-oriented computer language designed to solve arithmetic and logical programs.

full range floating zero - A characteristic of a numerical machine tool control permitting the zero point on an axis to be readily shifted over a specified range. The control retains information on the location of "'permanent" zero.

G code - A word addressed by the letter G and followed by a numerical code, defining preparatory functions or cycle types in a CNC system. It presets certain states of the control.

gage height - A predetermined partial retraction point along the Z axis to which the cutter retreats from time to time to allow safe XY table travel.

hard copy - Any form of computer-produced printed document. Also, sometimes punch cards or paper tape.

hardware - Physical equipment.

head - A unit, usually a small electromagnet on a storage medium such as magnetic tape or a magnetic drum, that reads, records, or erases information on that medium. The block assembly and perforating or reading fingers used for punching or reading holes in paper tape.

helical interpolation – A form of interpolation that combines the circular interpolation for two axes with linear movement of a third axis in the current plane; this permits the definition of a helical path in three-dimensional space.

IC - (see integrated circuit)

incremental dimension - A dimension expressed with respect to the preceding point (last point) in a sequence of points.

incremental system - Control system in which each coordinate or positional dimension is taken from the previous position, Fig. G19.

indexing - Movement of one axis at a time to a precise point from numeric commands.

inhibit - To prevent an action or acceptance of data by applying an appropriate signal to the appropriate input.

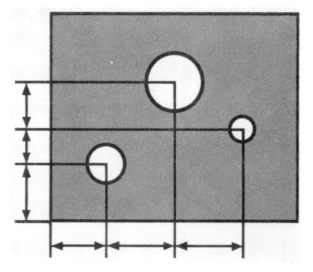

Fig. G19 Incremental system. *(Cincinnati Milacron, Inc.)*

input/output (I/O) devices – Computer equipment used to provide data to the computer (input devices such as a keyboard) or to people or other machines (output devices such as a display or printer, Fig. G20.
- The I/O ports receive information such as 2+6, and sends it to the CPU.
- The CPU stores the 2 and 6 in the RAM and the + in the ROM.
- The CPU processes the parts and sends the end product 8 back through the I/O ports.

instruction - A statement that specifies an operation and the values or locations of its operands.

integrated circuit (IC) – A combination of interconnected passive and active circuit elements incorporated on a continuous substrate. (See chip).

interface - 1. A hardware component or circuit linking two pieces of electrical equipment having separate functions, e.g., tape reader to data processor or control system to machine. 2. A hardware component or circuit for linking the computer to an external input/output (I/O) device.

interpolation – CNC routines produced by the interpolation module in the MCU to calculate the intermediate points the cutter must follow in order to generate a particular mathematically defined or approximated path, Fig. G21.
- The most common interpolation systems are linear interpolation, circular interpolation, helical interpolation, and NURBS interpolation.
- Each of these systems allows the programmer or operator to generate machine instructions for linear or curvilinear paths, using relatively few input parameters.

jog - A function that provides a controlled feed rate for manual movement of the axes, usually during setup.

leading zeros - Redundant zeros to the left of a decimal point or number.

linear interpolation - A function of a control whereby data points are generated between

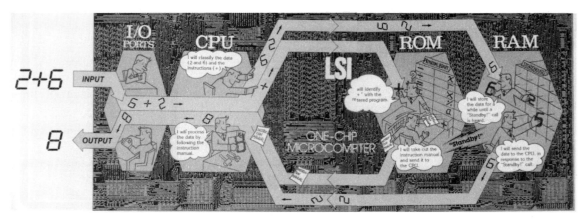

Fig. G20 Input/output devices. *(Sharp Electronics)*

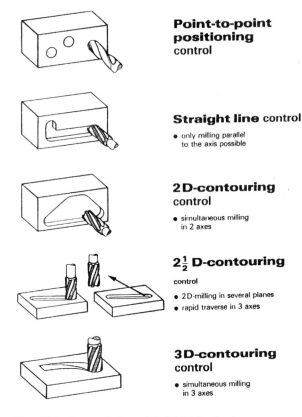

Point-to-point positioning control

Straight line control
- only milling parallel to the axis possible

2D-contouring control
- simultaneous milling in 2 axes

2½ D-contouring control
- 2D-milling in several planes
- rapid traverse in 3 axes

3D-contouring control
- simultaneous milling in 3 axes

Fig. G21   Interpolation. *(Deckel-Maho, Inc.)*

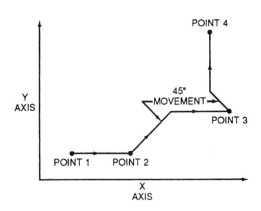

Fig. G22   Linear interpolation. *(Kelmar Associates)*

given coordinate positions to allow simultaneous movement of two or more axes of motion in a linear (straight line) path, Fig. G22.

logic -
1. Electronic devices used to govern a particular sequence of operations in a given system.
2. Interrelation or sequence of facts or events when seen as inevitable or predictable.

loop – A structure of computer or part program language in which a sequence of instructions is executed repeatedly until a terminating condition is satisfied.

LSI - (see large-scale integration)

machine control unit (MCU) – The actual CNC system control that reads and interprets the program of instructions and convert it into mechanical actions of the machine tool or other processing equipment.

machining center - A CNC machine tool capable of automatically drilling, reaming, tapping, milling, and boring multiple faces of a part and often equipped with a system for automatically changing cutting tools.

macro (command/subprogram) – The ability of a CNC programming language that allows a sequence of commands, in the form of a variable type of subprogram, to produce a series of tool paths within the part program.

management information system (MIS) - An information feedback system that transmits data from the machine to management, implemented by a computer.

manual data input (MDI) – The manual programming of part programs to the memory of a CNC control unit by using the console keyboard at the site of the machine tool. Communication between the operator-programmer and the system is through the CRT display monitor and the keyboard.

manual part programming - The manual preparation of a manuscript in machine control language and format to define a sequence of commands for use on a CNC machine.

manuscript - Form used by a part programmer for listing detailed manual or computer part programming instructions, Fig. G23.

MDI - (see manual data input)

memory - A device or medium used to store information in a form that can be understood by the computer hardware.

microprogramming - A programming technique in which multiple-instruction operations can

CNC Program Worksheet

Part # <u>CNC-133A</u>
Program # <u>06789</u>

N	G	X (J) (D)	Y (K) (S)	Z	F (L) (T) (H)	M	Remarks
010	20	G90					
020	00	G54 X·500 Y·560				M03	
030	43			Z·100	H01	M08	
040	01		S2000	Z-·125	F5·0		
050			Y·500				
060		X1·500					
070			Y1·000				
080		X1·060					
090	00			Z·100		M09	
100	28			Z·100		M05	
110	28	X1·060	Y1·060				
120						M30	
%							REWIND CODE

Fig. G23   Programming manuscript. *(Kelmar Associates)*

be combined for greater speed and more efficient memory use.

microsecond - One millionth of a second.

millisecond - One thousandth of a second.

mirror imaging – A CNC feature that reverses the sign of programmed dimensions in one or two axes by means of a switch, allowing opposite-hand geometry to be produced by a single programmed command, Fig. G24.

• It can also be used to repeat geometric features, programmed in a single quadrant, in other quadrants.

MIS - (see management information system)

miscellaneous function (M code) – Sometimes called a machine function because it is related to performing some type of operation on a CNC machine such as controlling the spindle, stopping programs, turning coolant ON or OFF, etc.

modal - A set of commands retained in a system until a new command cancels or replaces them.

nanosecond - A billionth of a second; a common unit of measure of computer operating speed.

numerical control (NC) - A technique of operating machine tools or similar equipment in which motion is developed in response to numerically coded commands.

off-line - Operating software or hardware not under the direct control of a central processor, or operations performed while a computer is not monitoring or controlling processes or equipment.

offset - The steady-state deviation of the controlled variable from a fixed setpoint.

on-line - A condition in which equipment or programs are under direct control of a central processor.

open-loop system - A control system that has no way to compare the output with the input or no feedback for control purposes, Fig. G25.

optical encoder – A feedback sensor device used in CNC that measures linear or rotary motion

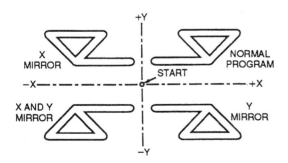

Fig. G24   Mirror imaging. *(Superior Electric Co.)*

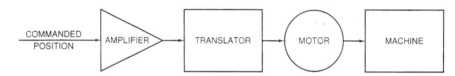

Fig. G25 Open-loop system. *(Allen Bradley Co.)*

by detecting the movement of markings past a fixed beam of light.

optional stop - A Miscellaneous Function command similar to Program Stop except that the control ignores the command unless the operator has previously pushed a button to validate the command (M01).

output - Dependent variable signal produced by a transmitter, control unit, or other device.

overshoot - The amount that a controlled variable exceeds its desired value after a change of input.

parabola - A plane curve generated by a point moving so that its distance from a fixed second point is equal to its distance from a fixed line.

parity check - A test of whether the number of 1s or 0s in an array of binary digits is odd or even to detect errors in a group of bits.

part program - Specific and complete set of data and instructions written in source languages for computer processing or written in machine language for manual programming for the purpose of manufacturing a part on a CNC machine.

part programmer - A person who prepares the planned sequence of events for the operation of a CNC machine tool.

part programming using CAD/CAM – An advanced form of computer-assisted part programming that uses an interactive graphics system equipped with CNC programming software, Fig. G26.

- The actions indicated by the commands are displayed on the graphics monitor, providing visual feedback to the programmer.
- Certain portions of the programming cycle are automated by the CNC programming software to reduce the total programming time required.
- Advanced CAD/CAM systems have the capability to automate portions of tasks of both geometry definition and tool path specification.

plotter – An output device that produces a hard copy record of data, such as cutter path, in the form of a two-dimensional graphic drawing.

picosecond - One millionth of one microsecond.

point-to-point control system - A CNC system that controls motion only to reach a given end point but exercises no path control during the transition from one end point to the next.

polar coordinates – A system of coordinates that locates a point plane with respect to its dis-

Fig. G26 Part programming CAD/CAM.
(AMT—The Association for Manufacturing Technology)

tance from a fixed point (origin or pole), and the angle this line makes starting from the polar reference line, Fig. G27.

positioning/contouring - A type of CNC system that has the capability of contouring, without buffer storage, in two axes, and the ability of positioning in a third axis for such operations as drilling, tapping, boring, etc.

postprocessor - The part of the software that converts all the cutter path coordinate data (obtained from the general-purpose processor and all other programming instructions and specifications for the particular machine and control) into a form which the machine control can interpret correctly.

preparatory function (G-code) - A CNC program command that changes the mode of operation of the control (generally noted at the beginning of a block by "G" plus two digits).

printed circuit - A circuit for electronic components made by depositing conductive material in continuous paths from terminal to terminal on an insulating surface.

program -

1. A plan for the solution of a problem. A complete program includes plans for the transcription of data, coding for the computer, and absorption of the results into the system. The list of coded instructions is called a *routine*.

2. To plan a computation or process from the asking of a question to the delivery of the results, including the integration of the operation into an existing system. Thus, programming consists of planning and coding, including numerical analysis, systems analysis, specification of printing formats, and any other functions necessary to the integration of a computer in a system.

programmed dwell - The capability of commanding delays in program execution for a programmable length of time.

program stop - A Miscellaneous Function (M00) command to stop the spindle, coolant, and feed after completion of the dimensional move commanded in the block. To continue with the remainder of the program, the operator must restart the program.

punch card - A piece of lightweight cardboard on which information is represented by holes punched in specific positions. (now obsolete)

punched tape - A rarely-used storage medium made of paper, plastic, and polyester laminates, that is used for the permanent storage and loading of CNC part programs, and on which characters are represented by combination of holes, Fig. G28.

- Punched tape for CNC applications is 1.000 in.(25 mm) wide, has eight parallel tracks of holes along its length, and holds 10 characters per 1.000 in. (25 mm) of length.

- The presence or absence of a hole in a certain position represents bit information, and the entire collection of holes constitutes the CNC program.

pulse - A short-duration change in the level of a variable.

quadrant - Any of the four parts into which a plane is divided by rectangular coordinate axes lying in that plane.

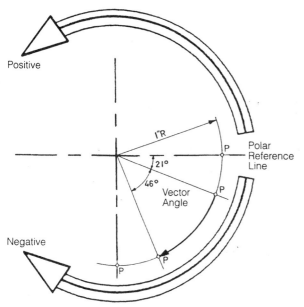

Fig. G27 Polar coordinates. *(Kelmar Associates)*

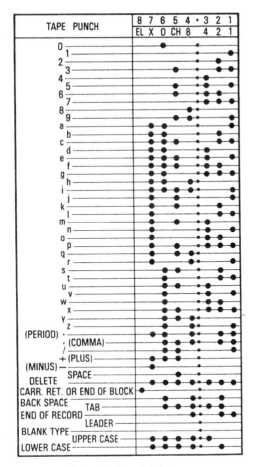

Fig. G28 Punched tape.

random-access memory (RAM) - A form of temporary internal storage contents that can be recalled and changed by the user; also known as read-and-write memory.

reader - A device capable of sensing information stored in off-line memory media (cards, paper tape, magnetic tape, etc.) and generating equivalent information in an on-line memory device (e.g., register or memory locations).

read-only memory (ROM) - Permanent internal memory containing data or operating instructions that can be recalled but not changed by the user.

real-time clock - The circuitry that maintains time for use in program execution and event initiation.

reference block - A block within a CNC program identified by an O or H in place of the word

address N that contains sufficient data to enable resumption of the program following an interruption. (This block should be located at a convenient point in the program to enable the operator to reset and resume operation.)

resolver -
1. A mechanical-to-electrical transducer whose input is a vector quantity and whose outputs are components of the vector.
2. A transformer whose coupling may be varied by rotating one set of windings relative to another. It consists of a stator and rotor, each having two distributed windings 90 electrical degrees apart.

right hand rule – A method of determining the primary axes of a CNC milling machine where the thumb indicates the X axis, the index finger shows the Y axis, and the middle finger the Z axis. (See Fig. G12)

ROM - (see read-only memory)

rotary axes – The axes A, B, and C that define the rotary movements around the axes parallel to X, Y, and Z. Positive rotation is clockwise, negative is counterclockwise, Fig. G29.

routine - A series of computer instructions that performs a specific task.

scale - To change a quantity by a given factor, to bring its range within prescribed limits.

scanner - The equipment used to digitize coordinate information from a master and convert it

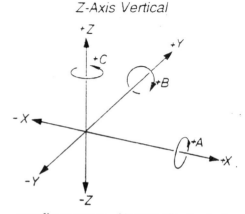

coordinate system. *(Courtesy Allen Bradley.)*

Fig. G29 Rotary axes. *(Allen Bradley Co.)*

to the computer program for later recreation of the master shape on a CNC machine.

sequence number (N-word) – Positive integer number preceded by the letter *N* that identifies the separate blocks, and the relative positions of those blocks within a CNC part program.

servo amplifier - The part of the servo system that increases the error signal and provides the power to drive the machine slides or the servo valve controlling a hydraulic drive.

setpoint - The position established by an operator as the starting point for the program on a CNC machine.

shaft encoder – A transducer that converts the angular position of a rotating shaft into digital coded form using a digitally coded disk, Fig. G30. See also Optical Encoder.

significant digit - A digit that contributes to the precision of a numeral. The number of significant digits is counted beginning with the digit contributing the most value, called the most significant digit, and ending with the one contributing the least value, called the least significant digit.

software - The collection of programs, routines, and documents that directs the operation of a computer.

storage - A memory device in which data can be entered and held and from which it can be retrieved.

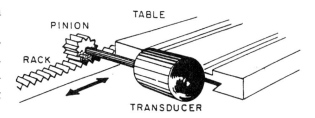

Fig. G30 **Shaft encoder.** *(Kelmar Associates)*

subroutine/subprogram – A separately defined part of a computer or CNC part program that can be called to execute from various points in the main program, Fig. G31.

- It is used to simplify and shorten programming by defining commonly used sequences once only, and calling them into the program as required with a program number.

system software – The programs that direct the internal operations of the computer while it is executing instructions from the user. They include functions such as translating languages, managing computer resources, and developing programs.

tape - A magnetic or perforated paper medium for storing information.

tool changer – Mechanisms that automatically change cutting tools on a CNC machine under program control.

- It may take the form of a carousel with a variety of cutting tools for machining cen-

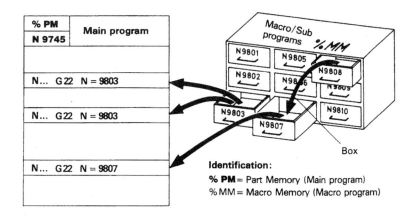

Fig. G31 **Subroutines.** *(Deckel-Maho Inc.)*

ters, or a special device at the end of a robot arm that provides for quick changes of the end-effector or tool.

tool function - A program command identifying a tool and calling for its selection. The address is normally a T word.

tool length compensation - A manual input means that eliminates the need for preset tooling and allows the programmer to program all tools as if they were of equal length.

tool nose radius compensation (TNRC) – Tool offset values that allow for small variations in tool point locations on CNC lathes. Although the tool point exists as a programmable point, it may not exist physically.

tool offset -
1. A correction for tool position parallel to a controlled axis.
2. The ability to reset tool position manually to compensate for tool wear, finish cuts, and tool exchange.

tool path - The center line of a CNC cutting tool while a cutting operation such as milling, drilling, or boring is performed.

trailing zero suppression – Redundant zeros to the right of the decimal point or number.

transducer – A unit that converts energy from one form to another such as temperature, pressure, and weight into electrical signals. Transducers are the basic mechanisms used to provide sensing functions for machine tools.

turning center - A lathe-type CNC machine tool capable of automatically boring, turning outer and inner diameters, threading, and facing multiple diameters and faces of a part, and often equipped with a system for automatically changing or indexing cutting tools, Fig. G32.

turnkey system - A term applied to an agreement whereby a supplier will install a CNC or computer system and have total responsibility for building, installing, and testing the system.

USACII - (see American Standard Code for Information Interchange)

vector - A quantity that has magnitude, direction, and sense, and that is commonly represented

Fig. G32 Turning Center.
(Cincinnati Machine, a UNOVA Co.)

by a directed line segment whose length represents the magnitude and whose orientation in space represents the direction.

vector feed rate - The resultant feed rate at which a cutter or tool moves with respect to the work surface. The individual slides may move slower or faster than the programmed rate, but the resultant movement is equal to the programmed rate.

virtual reality – An interactive 3D simulation process that allows a person to analyze the operation of a CNC machine tool, its controller, and the material-removal process on a computer.

word address format – A CNC part program format in which each word is preceded by a letter to identify its function and to address the data to a particular location in the controller unit, Fig. G33.
- The X prefix identifies an X-coordinate word; the S prefix identifies spindle speed, and so on.
- The standard sequence of words for a two-axis CNC system is N-word, G-word, X-word, Y-word, F-word, S-word, T-word, M-word, and EOB.

word length - The number of bits or characters in a word.

X axis - Axis of motion that is always horizontal and parallel to the workholding surface.

Program

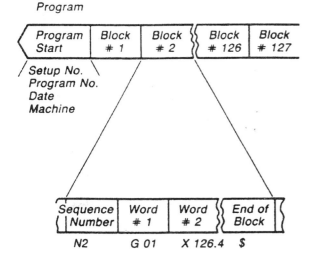

Fig. G33 Word address format. *(Allen Bradley Co.)*

Y axis - The axis of motion that is perpendicular to both X and Z axes.

Z axis - The axis of motion that is always parallel to the principal spindle of the machine.

zero offset - A characteristic of a numerical machine tool control permitting the zero point on an axis to be shifted readily over a specified range. The control retains information on the location of the "permanent" zero.

zero shift - Same as zero offset, except the control does not retain information on the location of the "permanent" zero.

zero suppression - The elimination of insignificant zeros to the left of significant digits, usually before printing.

Index

How to use this index. The text is arranged in numbered sections, with sequentially-numbered units, each of which starts a new set of page numbers. In the index, separated by hyphens, the section number is given first, the unit number second, and the page number third. Thus, 7-14-4 means Section 7, Unit 14, page 4, and so on. Where no page number is given, reference is to the entire unit, where no unit number is given, reference is to the entire section.